T0309388

# Breaking Paradigms
## in Atomic and Molecular Physics

# Breaking Paradigms
## in Atomic and Molecular Physics

### Eugene Oks
Auburn University, USA

 **World Scientific**

NEW JERSEY · LONDON · SINGAPORE · BEIJING · SHANGHAI · HONG KONG · TAIPEI · CHENNAI

*Published by*

World Scientific Publishing Co. Pte. Ltd.
5 Toh Tuck Link, Singapore 596224
*USA office:* 27 Warren Street, Suite 401-402, Hackensack, NJ 07601
*UK office:* 57 Shelton Street, Covent Garden, London WC2H 9HE

**British Library Cataloguing-in-Publication Data**
A catalogue record for this book is available from the British Library.

ISBN 978-981-4619-92-9

In-house Editor: Song Yu

Typeset by Stallion Press
Email: enquiries@stallionpress.com

Printed in Singapore

With love to
*Alex, Ellen, and Andrew*

# Contents

*"All truth goes through three steps. First, it is ridiculed. Second, it is violently opposed. Finally, it is accepted as self-evident."*

Arthur Schopenhouer

# Chapter 1

# Introduction

In this book, we present a number of counterintuitive theoretical results. These fundamental results break several paradigms of quantum mechanics and provide alternative interpretations of some important phenomena in atomic and molecular physics.

In Chapter 2, we revisit singular solutions of the Schrödinger and Dirac equations — the solutions that were always rejected. We demonstrate that they should not have been always rejected: They can be legitimate and necessary for explaining some experimental results. Specifically, they resolve a long-standing dispute concerning the high-energy tail of the linear momentum distribution in the ground state of hydrogen atoms/hydrogen-like ions. Apart from the fundamental importance, these results also have a practical importance. Namely, they open up a unique way to test intimate details of the nuclear structure by performing atomic (rather than nuclear) experiments and calculations. The results presented in Chapter 2 have been published in paper [1.1].

In Chapter 3, we take a fresh look at one of the most important processes in atomic and molecular physics: Charge exchange. Charge exchange was always considered as an inherently quantal phenomenon — see, e.g., [1.2, 1.3]. We demonstrate that charge exchange is not really an inherently quantal phenomenon, but rather has classical roots. Charge exchange and crossings of corresponding energy levels that enhance charge exchange are strongly connected with problems of energy losses and of diagnostics in high temperature plasmas. Besides, charge exchange was proposed as one of the most effective mechanisms for population inversion in the soft X-ray and VUV ranges.

One of the most fundamental theoretical playgrounds for studying charge exchange is the problem of electron terms in the field of two stationary Coulomb centers (TCC) of charges Z and Z' separated by a distance R. It presents fascinating atomic physics: The terms can have crossings and quasicrossings [1.4]. These rich features of the TCC problem also manifest in a different area of physics such as plasma spectroscopy: A quasicrossing of the TCC terms, by enhancing charge exchange, can result in an unusual structure (a dip) in the spectral line profile emitted by a Z-ion from a plasma consisting of both Z- and Z'-ions, as was shown theoretically and experimentally. The paradigm was that the above sophisticated features of the TCC problem and its flourishing applications are inherently quantum phenomena. We disprove this paradigm by presenting a purely classical description of the crossings of energy levels in the TCC problem leading to charge exchange.

The classical TCC systems represent one-electron diatomic Rydberg quasimolecules encountered in plasmas containing more than one sort of ions. We also study how their classical energy terms are affected by magnetic or electric fields, or by the screening by plasma electrons.

Apart from the fundamental importance, these results also have a practical importance. We apply them to the problem of continuum lowering in plasmas, which plays a key role in calculations of the equation of state, partition function, bound-free opacities, and other collisional atomic transitions in plasmas.

There are also two additional classical studies using (as the starting point) the same formalism as above, though not dealing with crossings of energy terms: Muonic-electronic negative hydrogen ion, and diatomic Rydberg quasimolecules in a laser field. These studies are presented in Appendices A and B, respectively.

The results presented in Chapter 3 have been published in papers [1.5–1.13].

In Chapter 4, we turn to the most challenging problem of classical physics that led to the development of quantum mechanics: The failure of classical physics to explain the stability of atoms. Classically, electrons revolving around a nucleus should emit electromagnetic radiation, leading to the loss of the energy and the fall to the nucleus — see, e.g., [1.14]. We show that this challenging problem can be actually solved within a classical formalism from first principles: The fall of atomic electrons on the nucleus due to the radiative loss of the energy, which seemed to be classically unavoidable, does not occur within Dirac's generalized Hamiltonian dynamics (GHD) applied to atomic physics.

The GHD is a purely classical formalism for systems having constraints; it incorporates the constraints into the Hamiltonian. Dirac designed the GHD specifically for applications to quantum field theory [1.15–1.18]. In our application of the GHD to atomic and molecular physics, integrals of the motion are chosen as the constraints. The result is the appearance of classical non-radiating states coinciding with the corresponding quantal stationary states. The underlying physics can be interpreted as a non-Einsteinian time dilation. As an application, we discuss advantages of this formalism over classical and semiclassical models employed in chemical physics for describing electronic degrees of freedom. The results presented in Chapter 4 have been published in papers [1.19, 1.20].

In Chapter 5, we present counterintuitive results concerning the role of the spin–spin interaction in two-electron atoms/ions. Usually this interaction was considered as an unimportant correction to the binding energy. However, it turns out that the spin–spin interaction actually makes a significant contribution to the binding energy if the singular nature of this interaction is properly taken into account. For the antiparallel orientation of spins, the allowance is made for the finite nature of motion of the electrons in the region of very small interelectronic distances determined by the change of the configuration space by the magnetic fields of the spins. The paired state formed in this case is described by the one-electron wavefunction for a particle with double mass and charge and principal quantum number $n = 2$. As a result, a good agreement with experimental values of the ionization potential is obtained for a wide range of two-electron atoms/ions without resorting to variational procedures. The results presented in Chapter 5 have been published in paper [1.21].

In Chapter 6, we revisit the following problem. An isolated hydrogen atom emits practically infinite series of spectral lines, corresponding to radiative transitions from the upper level of the principal quantum number n to the lower level of the principal quantum number $n_0$. (Here by "hydrogen atoms" and "hydrogen spectral lines" we mean atoms and spectral lines of hydrogen, deuterium, and tritium.) In other words, there is practically no restriction on how high the number $n$ can be. However, when a hydrogen atom is placed in a plasma, the observed spectral series (such as, e.g., Balmer or Paschen series) terminates at some $n = n_{max}$. According to the Inglis–Teller concept, developed as early as in 1939 [1.22] and widely used for diagnostics of laboratory and astrophysical plasmas, $n_{max}$ is controlled by the electron density $N_e$ and thus can serve for measuring $N_e$.

However, Inglis–Teller concept breaks down in plasmas with sufficiently large magnetic field B. We show that in this case, $n_{max}$ is controlled by the product $BT^{1/2}$, where $T$ is the atomic temperature. This fundamental result has an important practical application. Namely, in magnetized plasmas, the number $n_{max}$ of the last hydrogen line observed in a particular experiment, can be used for measuring the atomic temperature $T$, if the magnetic field is known, or the magnetic field $B$, if the temperature is known. We present examples of applications to the edge plasmas of tokamaks and to solar chromosphere. (Tokamaks are a type of plasma machines designed for the research in the area of magnetically-controlled thermonuclear fusion leading to a practically-inexhaustible source of energy.) The results presented in Chapter 6 have been published in paper [1.23].

In Chapter 7 containing conclusions, we summarize the presented counterintuitive results. We especially discuss the role and importance of classical models that provide an adequate picture of the reality.

## Chapter 2

# Role of Singular Solutions of Quantal Equations in Atomic Physics

## 2.1. A Long-Standing Mystery of the High-Energy Tail of the Linear Momentum Distribution in the Ground State of Hydrogen Atoms or Hydrogen-Like Ions (GSHA)

Tests of fundamentals of quantum theory had long ago passed the level of non-relativistic quantum mechanics and had moved three levels higher (passing quantum electrodynamics and then quantum chromodynamics) to testing what might occur beyond the so-called standard model. However, at the level of non-relativistic quantum mechanics, there still remains a *fundamental dispute* started over 35 years ago.

The problem we are talking about is the distribution function $f(p)$ of the linear momentum $p$ in GSHA. Hydrogen atoms, being the simplest atoms, were always considered as a test-bench for checking atomic theories versus experiments.

In 1935 Fock [2.1] derived non-relativistic wave functions for a hydrogen atom or a hydrogen-like ion in the momentum representation (here and below, for brevity we use simply the word "momentum" meaning "linear momentum"). As a particular result for the bound electron in the GSHA, he obtained a distribution $dw = f_{\text{quant}}(p)\,dp$ with the following distribution function $f(p)$:

$$f_{\text{quant}}(p) = 32p^2 p_0^5 / [\pi (p_0^2 + p^2)^4], \quad p_0 \equiv Zme^2/\hbar. \quad (2.1)$$

Here $Z$ is the nuclear charge; $p_0/Z \approx 1.992 \times 10^{-19}$ g·cm/s practically coincides with the atomic unit of the linear momentum; $m$ is the

5

reduced mass of an electron in a hydrogen atom or hydrogen-like ion. From Eq. (2.1) it follows that the high-energy tail of the momentum distribution (HTMD) has the form

$$f_{\text{quant}}^{\text{As}}(p) \equiv f_{\text{quant}}(p \gg p_0) \propto 1/p^6. \tag{2.2}$$

However, in the mid-60s, Gryzinski developed a classical "free-fall" model of hydrogen atoms [2.2] and then its modified version [2.3] (some authors later on suggested another advanced free-fall model [2.4]). In these models, the atomic electron moves along straight lines toward or away from the nucleus (with a possible exception of a small vicinity around the nucleus), so that the angular momentum of the electron is zero. The HTMD, corresponding to the free-fall models, has the form

$$f_{\text{class}}^{\text{As}}(p) \propto 1/p^4, \tag{2.3}$$

which is very different from the quantum HTMD given by Eq. (2.2) — see also a later paper [2.5] comparing the quantal and classical distribution functions $f(p)$ in detail. Based on the free-fall model, Gryzinski derived an amazing amount of well-working relations that yielded a very good agreement with experiments for a great variety of collisional processes between atoms (including hydrogen atoms) and charged particles (electrons and protons).

To avoid any confusion, we should note the following. Before developing the free-fall model, Gryzinski tried to eliminate a discrepancy between classical and quantal ionization cross-sections (IC) at *high incident energies.* At that time it was known that for high incident energies $E$, the classical IC falls of as $1/E$, while the quantal IC falls off as $\log(E)/E$ (see, e.g., [2.6]). Gryzinski has shown [2.7, 2.8] that it is possible to obtain a classical IC which falls off as $\log(E)/E$ if the HTMD for the bound electron would be $f_{\text{hyp}}^{\text{As}}(p) \propto 1/p^k$, $k = 3$ (here the suffix "hyp" stands for "hypothetical"). He also mentioned in [2.8] that due to the approximate character of the theory, a value of $k$ slightly different from 3 cannot be excluded. In 1970 Percival and Richards [2.9] showed that a classical IC, calculated by extending Bohr's correspondence principle, falls off as $\log(E)/E$ regardless of the HTMD of the bound electrons (for the latest development of the quantal-classical correspondence see a paper by Flannery and Vrinceanu [2.10]). Besides, some additional approximations of a secondary importance, used by Gryzinski in [2.8] and in his preceding papers, were also criticized later on; corrections, extensions, and simpler derivations were provided by a number of authors for some of the relations he derived (see, e.g., reviews [2.11–2.13]).

However, the above criticism does not relate to purely classical calculations based on the free-fall model. The free-fall model was successfully applied to various inelastic processes not only in hydrogen atoms, but in many other atoms and molecules as well — see, e.g., review [2.14] and references in the later paper [2.15].)

Moreover, attempts by the Percival's group [2.16, 2.17] to use in classical calculations for the bound electrons the quantal distribution function from Eq. (1), having the HTMD $\sim 1/p^6$, resulted in about 60% discrepancy with the experimental IC of atomic hydrogen by electrons at relatively *low incident energies* (LIE), while the employment of the free-fall model (where the HTMD $\sim 1/p^4$) for calculating the same IC [2.18–2.20] yielded a very good agreement with the experiments at the same range of energies. (At high incident energies, ICs calculated using either the HTMD from (2.2) or the HTMD from (2.3) agree well with each other and with the experiments.)

Later Kim and Rudd [2.21] introduced a modified classical binary-encounter model and achieved an agreement (as the Gryzinski's model did) with the LIE part of the experimental IC of atomic hydrogen by electrons. However, their model deals only with the averaged kinetic energy of the bound electron and therefore cannot distinguish between the two disputed forms of the HTMD. Besides, even if Kim and Rudd would have allowed for details of the HTMD, this would not have clarified the above dispute because the model from [2.21] has two other weak points. First, Kim and Rudd omitted one out of two terms which could be sensitive to the HTMD of the bound electrons. Second, they chose an approximate coefficient in front of the remaining term. Thus, in frames of the classical or semiclassical approaches, the above dispute remains unresolved up to now.

Concerning purely-quantal (hereafter called "quantal") calculations of the same IC, the situation is as follows. Early (perturbative) quantal calculations of the same IC showed an even much more dramatic disagreement with the experiments than the mixed classical-quantal calculations from [2.16, 2.17]: Various such calculations exceed the corresponding experimental results *by several times* at the LIE (see, e.g., Fig. 6 from review [2.14]). More recent (non-perturbative) quantal calculations by a distorted wave method [2.22] or by close coupling methods [2.23, 2.24] (as well as by a closely related method using $R$-matrix with pseudo-states [2.25]) improved the situation. However, different versions of the distorted wave method yield different results [2.26]. Moreover, even within the same version there is a significant ambiguity, leading to differences up to 50–100% at the incident energies $\leq 20$ eV — compare, e.g., Fig. 1 from [2.21] and Fig. 1 from [2.26]. As for the close coupling/$R$-matrix methods, their good agreement with one out of two sets of the experimental IC might be overshadowed by (and questioned due to) a dramatic disagreement — up to a factor of 2 — between the results of these methods for the width of the spectral line B II 2s–2p [2.27] and the corresponding benchmark experiments [2.28], while the latter is in a good agreement with semiclassical calculations [2.29]. We also note a hidden crossing theory [2.30], which reproduced very well the shape of the experimental IC at the LIE, but that did not reproduce well the absolute values of the experimental IC at the LIE.

The above does not diminish the ingenuity of the authors of [2.21–2.27, 2.30], but rather illustrates a well-known fact: The lower the incident electron energy, the stronger the electron correlations become, creating greater and greater computational difficulties for quantal calculations. Besides, it is unclear whether these non-perturbative quantal methods are sensitive enough to the HTMD.

So, one point we are trying to make is that the above fundamental dispute still remains unresolved: The experiments seem to favor a HTMD $\sim 1/p^k$, where $k$ is *at least* 1.5 *times smaller than in the quantum HTMD*. This puzzle is even more mind-boggling due to the following two facts. First, while the HTMD (2.3) corresponds to relatively large values of the linear momentum $p \gg p_0$, these values are still below the relativistic domain of $p$. Indeed, since for hydrogen atoms we have $p_0/mc = e^2/\hbar c \equiv \alpha \approx 1/137$, there is a significant range of $p$, where it seems that the non-relativistic quantum theory (used for deriving Eqs. (1) and (2)) should remain valid. Second, a non-relativistic classical treatment of a hydrogen atom can determine *exactly* both the energies and the wave functions, so that the dynamics of hydrogen atoms can be fully described in terms of the corresponding classical orbits, as was shown by Kay [2.31].

Another point is that a possible resolution of the above dispute might be connected to the problem of *singular solutions* of quantal equations, as will be shown below. A role of singular solutions of the Schrödinger or Dirac equations is a *fundamental problem in its own right*. With respect to the Dirac equation, such a study started in 1945, when Pomeranchuk and Smorodinskii [2.32] introduced a finite nuclear size into the Coulomb problem. This activity was resumed about 40 years ago — see, e.g., papers [2.33–2.35], as well as textbooks [2.36, 2.37]. In these studies, the interaction potential inside the nucleus was modeled either as a constant potential inside a charged spherical shell or as a potential of a uniformly charged sphere. All the preceding works focused primarily at the range of the nuclear charge $Z > 1/\alpha \approx 137$ — to determine the electrodynamic limit on the nuclear charge of stable hydrogen-like ions. It was shown that in the range of $Z > 1/\alpha$, for the above two model potentials inside the nucleus, it is possible to match the regular interior solution with both regular and *singular* exterior solutions. However, for $Z < 1/\alpha$, for the above two model potentials, it was found impossible to match the regular interior solution with the singular exterior solution — see, e.g., textbook [2.38]. As a result of these studies, the paradigm is that, even with the allowance for a finite nuclear size, singular solutions of the Dirac equation for the Coulomb problem should be rejected for $Z < 1/\alpha$.

Here we *break this paradigm*. First, we derive a general condition for matching a regular interior solution with a singular exterior solution of the Dirac equation for *arbitrary* interior and exterior potentials. Then we find explicit forms of several classes of potentials that allow such a match. Finally, we show that, as an outcome, the HTMD for the GSHA acquires terms falling off much slower than the $1/p^6$-law prescribed by the previously adopted quantal result.

## 2.2. Early, Unsuccessful Attempts to Explain the Mystery

For spherically-symmetric potentials, radial parts of non-relativistic wave functions both in the coordinate representation $R_{nl}(r)$ and in the momentum representation $P_{nl}(p)$ are interrelated as follows (see, e.g., [2.39])

$$P_{nl}(p) = [r/p]^{1/2}(i^{-l}/\hbar) \int_0^\infty dr J_{l+1/2}(pr/\hbar) R_{nl}(r) r, \qquad (2.4)$$

where $J_{l+1/2}(z)$ is the Bessel function.

For the GSHA, the non-relativistic quantal distribution function (2.1) is

$$f_{\text{quant}}(p) = |P_{10}(p)|^2 p^2,$$

$$P_{10}(p) = [2/(\pi)]^{1/2}(1/p) \int_0^\infty dr \sin(pr/\hbar) R_{10}(r) r, \qquad (2.5)$$

The set of equations (2.2) show: The fact that something could be wrong with the function $P_{10}(p)$ at large $p$ (at $p \gg p_0$) translates into a statement that *the radial part $R_{10}(r)$ of the corresponding coordinate wave function could be incorrect at small $r$* (namely, at $r \ll \hbar/p_0 = \hbar^2/(Zme^2)$, the latter quantity practically coinciding with the Bohr radius divided by $Z$).

For hydrogen atoms or hydrogen-like ions, the functions $R_{nl}(r)$ have the following behavior at small $r$ (see, e.g., [2.39, 2.40]) $R_{nl}(r) \propto r^l$, so that $R_{10}(r) \approx$ const at small $r$. Since for the GSHA, the experiments show that the true HTMD falls off much slower than the non-relativistic quantum HTMD (2.2), then a true radial part $R_{10}^{\text{true}}(r)$ of the coordinate wave function should have a relatively strong singularity at small $r$: $R_{10}^{\text{true}}(r) \propto 1/r^q$, $q \geq 1$.

Now we conduct a preliminary analysis that would help finding a resolution of this challenging fundamental enigma. As the first attempt, let us look more carefully at the textbook solutions of the Schrödinger equation for a hydrogen atom. After separating variables and reducing the problem to a one-dimensional equation for $R_{nl}(r)$, it turns out that the latter equation actually allows two classes of solutions characterized by two different types of behavior of $R_{nl}(r)$ at small $r$ — in addition to the regular $R_{nl}(r) \propto r^l$, it allows also a singular solution (see, e.g., [2.39, 2.40]): $R_{nl}(r) \propto 1/r^{l+1}$.

For $l \geq 1$ it is easy to reject the singular solution because the normalization integral $\int_0^\infty |R_{nl}(r)|^2 r^2 \, dr$ diverges at $r = 0$. As for $l = 0$, the normalization integral does not have any divergence at $r = 0$. The overwhelming majority of textbooks on quantum mechanics rejects the second option for $l = 0$ without any explanation. Only a couple of textbooks [2.39, 2.41] provide some explanations, but one of them is questionable while another explanation is incorrect.

Indeed, in the book [2.39], the singular solution for $l = 0$ is rejected on the ground that it would make the integral expressing the mean value of the kinetic energy diverge at $r = 0$. However, the original idea of Schrödinger was that limitations imposed on formal solutions of his equation should be only the requirements for the wave function to be single-valued and to allow the normalization (the latter requirement being equivalent to imposing boundary conditions) [2.42]. Therefore, an additional requirement for the mean kinetic energy to be finite lacks the elegance intrinsic to the Schrödinger's idea. Besides, this additional requirement implies that the mean kinetic energy of the *bound* electron is an observable quantity, what might be questionable.

As for the authors of the book [2.41], while "proving" that the singular solution for $l = 0$ cannot exist, they make an implicit assumption that such a solution can be expanded in the Laurent series at $r = 0$. However, only analytic functions can have the Laurent series expansion (see, e.g., [2.43]), while nothing prevents a formal solution to be a non-analytic function. Below we show that such a solution is indeed a non-analytic function, so that the "proof" in the book [2.41] is indeed incorrect.

An explicit form of the second formal solution of the Schrödinger equation for the GSHA for the entire range of $r$ was never published — to the best of our knowledge. By working out this problem in a straightforward way, we found the second formal solution to be

$$R_{10}(r) = \text{const}\{Ei(2r) - [\exp(2r)]/(2r)\}\exp(-r), \qquad (2.6)$$

where $Ei(z)$ is the exponential integral function (here and below we use the Coulomb units $Z = \hbar = m = e = 1$, unless $Z$, or $\hbar$, or $m$, or $e$ appears explicitly). At $r \to 0$, this solution has the following approximate form

$$R_{10}(r) \approx \text{const}[1/r + \log(1/r^2)]. \qquad (2.7)$$

Equations (2.6) and (2.7) show that this solution is indeed a non-analytic function at $r = 0$, but there is no reason whatsoever to reject this formal solution based on its behavior at $r = 0$. An actual reason for rejecting this solution comes from analyzing it at $r \to \infty$. It turns out that at $r \to \infty$ this solution behaves as $\exp(2r)$, thus making impossible to normalize it.

As the second attempt, let us move to relativistic quantum mechanics and look at the textbook solutions of the Dirac equation for a hydrogen atom. Similarly to the case of the Schrödinger equation, the Dirac equation for a hydrogen atom formally allows two classes of solutions characterized by two different types of behavior of the radial part $R_{n'j}(r)$ of the coordinate wave function at small $r$ (see, e.g., textbooks [2.36, 2.38, 2.44, 2.45])

$$R_{n'j}(r) \propto 1/r^{1+s}, \quad s = \pm[(j + 1/2)^2 - (Z\alpha)^2]^{1/2},$$
$$\alpha \equiv e^2/(\hbar c) \approx 1/137, \qquad (2.8)$$

where $n'$ is the radial quantum number and $j$ is the quantum number corresponding to the total (spin plus orbital) angular momentum. For the GSHA we have $n' = 0$ and $j = 1/2$, so that Eq. (2.2) reduces to: $R_{1/2}(r) \propto 1/r^q$, $q = 1 \pm [1 - (Z\alpha)^2]^{1/2}$.

We are interested only in the case of $Z\alpha \ll 1$. In this case, the solution corresponding to $k = 1 - [1 - (Z\alpha)^2]^{1/2} \approx (Z\alpha)^2/2 \approx 2.66 \times 10^{-5}Z^2$ is accepted, because its singularity at $r = 0$ is weak and does not prevent the normalization (below we refer to this solution as being "regular"). The solution corresponding to $q = 1 + [1 - (Z\alpha)^2]^{1/2} \approx 2$ is rejected because its strong singularity at $r = 0$ prevents the normalization.

## 2.3. Singular Solutions: General Results on Classes of Potentials to Which They are Applicable

Thus, for the GSHA, both the Schrödinger and the Dirac equations provide a formal solution having a relatively strong singularity at $r = 0$, which could be needed for resolving the dispute on the HTMD. However, these singular solutions cannot be normalized: For the Schrödinger equation it blows up at $r \rightarrow \infty$, while for the Dirac equation such a solution is "too much singular" at $r = 0$.

The above analysis did not take into account *a finite nuclear size*. In the previous studies that allowed for a finite nuclear size, the interaction potential inside the nucleus was modeled either as a constant potential inside a charged spherical shell or as a potential of a uniformly charged sphere — see, e.g., papers [2.32–2.35], as well as textbooks [2.36, 2.37]. All the preceding works focused primarily at the range of the nuclear charge $Z > 1/\alpha \approx 137$ — to determine the electrodynamic limit on the nuclear charge of stable hydrogen-like ions. It was shown that in the range of $Z > 1/\alpha$, for the two above model potentials inside the nucleus, it is possible to match the regular interior solution with both regular and *singular* exterior solutions. However, for $Z < 1/\alpha$, for the two above model potentials, it was found impossible to match the regular interior solution with the singular exterior solution — see, e.g., textbook [2.38]. As a result of these studies, the paradigm is that, even with the allowance for the finite nuclear size, singular solutions of the Dirac equation for the Coulomb problem should be rejected for $Z < 1/\alpha$.

Let us study the problem in the following, more general set-up. We consider an arbitrary spherically-symmetric interaction potential $V(r)$, which takes two different forms in the interior region $r < R$ and in the exterior region $r > R$. It is known from the previous studies [2.32–2.37] that the matching of the solutions of the Dirac equation at the boundary $r = R$ reduces to the requirement that the ratio $f(r)/g(r) \equiv \rho(r)$ of the two radial components of the Dirac bispinor should be continuous at $r = R$. Therefore, instead of using two coupled differential equations for $f(r)$ and $g(r)$, it would be more convenient to start directly from the following Ricatti-type equation for $\rho(r)$ (see Eq. (5.22) from [2.36]). For the ground state it reads

$$\rho' = -2\rho/r + [V(r) - E + 1] + [V(r) - E - 1]\rho^2, \qquad (2.9)$$

where $E$ is the total energy and $\rho' \equiv d\rho/dr$. Here and below we use the natural units $c = \hbar = m = 1$. Introducing $u(r) \equiv \rho(r)r^2$, we simplify Eq. (9) to the form:

$$u' = r^{-2}[V(r) - E - 1][u(r)]^2 + [V(r) - E + 1]r^2. \qquad (2.10)$$

We study solutions of Eqs. (2.9) and (2.10) at $r \ll 1$ for the case of $R \ll 1$. For a *regular* solution the interior region ($r < R$), from general properties of the solutions of the Dirac equation we should expect that $\rho_{\mathrm{reg}}(r) \ll 1$, so that $u_{\mathrm{reg}}(r)/r^2 \ll 1$. Therefore, we can disregard the first term in the right side of Eq. (2.10) and obtain:

$$u_{\mathrm{reg}}(r) \approx \int_0^r V(r')r'^2\,dr' + (1 - E)r^3/3. \qquad (2.11)$$

For a *singular* solution in the exterior region ($R < r \ll 1$), from general properties of the solutions of the Dirac equation we should expect that $\rho_{\mathrm{sing}}(r) \gg 1$, so that $u_{\mathrm{sing}}(r)/r^2 \gg 1$. Therefore, we can disregard the second term in the right side of Eq. (2.10) and obtain

$$du_{\mathrm{sing}}/u_{\mathrm{sing}}^2 \approx [V(r) - E - 1]/r^2, \qquad (2.12)$$

so that

$$u_{\mathrm{sing}}(r) \approx \left\{ \int_r^\infty [V(r')/r'^2]\,dr' - (1 + E)/r \right\}^{-1}. \qquad (2.13)$$

Thus, the matching condition $u_{\mathrm{reg}}(R) = u_{\mathrm{sing}}(R)$ takes the form:

$$\int_0^R V(r')r'^2\,dr' + (1 - E)r^3/3$$

$$\approx \left\{ \int_R^\infty [V(r')/r'^2]\,dr' - (1 + E)/r \right\}^{-1}. \qquad (2.14)$$

## 2.4. Engaging Singular Solutions for the Successful Explanation of the Mystery

Now we specifically study the GSHA, so that $V(r) = -Z\alpha/r$ at $r \geq R$ and also $(1 - E) \ll 1$. The latter condition allows disregarding the second term in the left side of Eq. (2.14). It is easy to verify that for the interior region $(r < R)$ indeed neither a constant interaction potential $V = -Z\alpha/R$, corresponding to a charged spherical shell, nor an interaction potential of a uniformly charged sphere $V = (r^2/R^2 - 3)Z\alpha/(2R)$ satisfy Eq. (2.14).

However, from experiments on the elastic scattering of electrons on protons it is well known that the charge density inside proton neither has a peak at the periphery nor is constant, but rather it has a maximum at $r = 0$ [2.46–2.48]. This means that for $r < R$, a more realistic interaction potential should rise toward $r = R$ much faster than for $V = (r^2/R^2 - 3)Z\alpha/(2R)$. *The central point is that such interaction potentials can satisfy Eq.* (2.14) *and thus provide a match of the singular solution for the Coulomb field in the exterior region with a regular interior solution.* We illustrate this point by the following two particular examples.

First, for the region $r \leq R$ we consider a class of interaction potentials of the form:

$$V(r) = -(Z\alpha/R)\exp[(R - r)/b], \quad 0 < b \ll R. \qquad (2.15)$$

For this class of potentials, Eq. (2.14) yields: $(2Z\alpha b^3/R)\exp(R/b) \approx 2/R + Z\alpha/(2R^2)$. The latter equation has, e.g., the following solutions:

(A) $R \approx -b\ln(Z\alpha b^3) \equiv R_1$, where $b \ll -1/\ln(Z\alpha)$ so that $Z\alpha \ll R_1 \ll 1$;

(B) $R \approx Z\alpha$, provided that $b \approx -Z\alpha/\ln(4Z^4\alpha^4/5)$.

Second, for the region $r \leq R$ we consider another class of interaction potentials:

$$V(r) = -(Z\alpha/R)(R^m + b^m)/(r^m + b^m), \quad m \geq 3, \quad 0 < b \ll R. \qquad (2.16)$$

For instance, for $m = 6$, Eq. (2.14) yields: $\pi Z\alpha R^5/(6b^3) \approx 2/R + Z\alpha/(2R^2)$. The latter equation has, e.g., the following solutions:

(A) $R \approx [12/(\pi Z\alpha)]^{1/6}b^{1/2} \equiv R_2$, where $b \ll 1/(Z\alpha)^{1/3}$, so that $Z\alpha \ll R_2 \ll 1$;

(B) $R \approx Z\alpha$, provided that $b \approx (\pi Z^7\alpha^7/15)^{1/3}$.

Thus, for those interaction potentials in the interior region that rise rapidly enough toward the boundary $r = R$, it is in fact possible to match the singular solution for the Coulomb field in the exterior region with a regular interior solution. Therefore, for $r \geq R$, the radial part of the Dirac bispinor for the GSHA is a linear combination of both the regular and singular solutions for the Coulomb field. Then, by using the well-known forms of these regular and singular solutions [2.34–2.36, 2.38, 2.45] and keeping only the leading terms in the pre-exponential factor for $R \leq r \ll 1/(Z\alpha)$, we obtain for the GSHA:

$$f(r) \approx -\{1/r^{1-\gamma} + [(Z\alpha)^2(2\lambda)^{2\gamma}]^{-1}2\Delta/r^{1+\gamma}\}$$
$$\times (2\lambda)^{1/2+\gamma}(1 - E)^{1/2}\exp(-\lambda r),$$
$$g(r) \approx \{1/r^{1-\gamma} + [(2\gamma - 1)(Z\alpha)^2(2\lambda)^{2\gamma}]^{-1}\Delta/r^{\gamma}\}$$
$$\times (2\lambda)^{1/2+\gamma}(1 + E)^{1/2}\exp(-\lambda r),$$
$$\gamma \equiv (1 - Z^2\alpha^2)^{1/2}, \quad \lambda \equiv (1 - E^2)^{1/2},$$
$$\Delta \equiv (E_0 - E), \quad E_0 \equiv (1 - Z^2\alpha^2)^{1/2}. \tag{2.17}$$

Here $E_0$ is the unperturbed energy of the GSHA, $E$ is the energy of the GSHA perturbed by a finite nuclear size.

Now we can calculate the HTMD $f_{\text{RQFN}}^{\text{As}}(p)$ for the solution from Eq. (2.17) for $p_0 \ll p \ll 1/R$ (here and below the suffix RQ stands for "relativistic quantal" and the suffix FN stands for "finite nucleus"). This can be done analytically by using a known integral

$$\int_0^\infty x^a[\exp(-\gamma r)]\sin(px)\,dx$$
$$= [\Gamma(1 + a)/(p^2 + \gamma^2)^{(1+a)/2}]\sin[(1 + a)\arctan(p/\gamma)], \tag{2.18}$$

where $\Gamma(z)$ is the gamma-function. As a result, for the GSHA we obtain:

$$f^{As}_{RQFN}(p) \propto [1/p^{6-\delta} + (2\delta^2)^{-1}\Delta/p^4 + \delta^{-3/2}\pi\Delta/p^3],$$
$$\delta \equiv (Z\alpha)^2 \ll 1, \quad \Delta < \delta^2. \tag{2.19}$$

Therefore, if we would approximate the HTMD (2.19), which is valid for $p_0 \ll p \ll 1/R$, by some effective power law $f^{As}(p) \propto 1/p^k$, we would get the value of $k$ noticeably smaller than 6.

We note that the mean kinetic energy has a finite value regardless of Eq. (2.19), since Eq. (2.19) is not valid in the ultrarelativistic range of $p \gg 1/R$. A finite mean kinetic energy is a general feature of interaction potentials having a finite value and a zero slope at the origin (see, e.g., [2.33]) and both classes of the interaction potentials (2.15) and (2.16) possess these properties.

## 2.5. Opening a Way to Test Intimate Details of the Nuclear Structure by Performing Atomic, Rather than Nuclear, Experiments

We *broke the paradigm* that, even with the allowance for the finite nuclear size, singular solutions of the Dirac equation for the Coulomb problem should be rejected for $Z < 1/\alpha$. We derived a general condition for matching a regular interior solution with a singular exterior solution of the Dirac equation for *arbitrary* interior and exterior potentials. We found explicit forms of several classes of potentials that allow such a match. We showed that, as an outcome, the HTMD for the GSHA acquires terms falling off much slower than the $1/p^6$-law prescribed by the previously adopted quantal result.

The above might be considered as a resolution of a long standing dispute between classical and quantal calculations of the HTMD for the GSHA in favor of the presented, more sophisticated quantal calculations. Besides, our results open up a *unique way to test intimate details of the nuclear structure by performing atomic (rather than nuclear) experiments* and calculations. This outcome is highly *counterintuitive*.

Indeed, the HTMD can be subdivided into three ranges: non-relativistic ($p_0 \ll p < 1$), relativistic ($1 < p < 1/R$), and ultrarelativistic ($1/R < p$). Obviously, modifications of the potential inside the nucleus (i.e., at $r < R$) affect the ultrarelativistic part of the HTMD ($p > 1/R$). However, we found that such modifications could also result in noticeable/*observable changes* of the HTMD not only in the adjacent relativistic range of the HTMD, but *in the non-relativistic part of the HTMD* as well.

Our results should not be perceived as casting any doubt on the computational methods used for quantal, fully-numerical calculations for electron ionization of hydrogen atoms. We also note that the modification of the wave function, predicted by Eq. (2.17) on the scale intermediate between the atomic and nuclear sizes, might be probed in quantal numerical calculations not only for electron ionization of hydrogen atoms, but for some other processes as well.

However, what we would really like to emphasize is that classical treatments of electron ionization should be sensitive to the modifications of the HTMD, predicted by Eq. (2.19). Therefore, we hope that our results could revitalize classical calculations of electron ionization at a more sophisticated level, which would rival computational methods used in quantal numerical calculations for LIE.

While resolving this long-standing dispute in favor of quantal calculations, our results do not diminish the significance of Gryzinski's works for atomic physics. His works inspired us to perform this study.

# Chapter 3

# Classical Description of Crossings of Energy Terms and of Charge Exchange

## 3.1. Brief History and Importance of the Corresponding Studies

Charge exchange and crossings of corresponding energy levels that enhance charge exchange are strongly connected with problems of energy losses and of diagnostics in high temperature plasmas — see, e.g., [3.1, 3.2] and references therein. Besides, charge exchange was proposed as one of the most effective mechanisms for population inversion in the soft X-ray and VUV ranges [3.3–3.6]. One of the most fundamental theoretical playgrounds for studying charge exchange is the problem of electron terms in the field of two stationary Coulomb centers (TCC) of charges $Z$ and $Z'$ separated by a distance $R$. It presents fascinating atomic physics: the terms can have crossings and quasicrossings.

The crossings are due to the fact that the well-known Neumann–Wigner general theorem on the impossibility of crossing of terms of the same symmetry [3.7] is not valid for the TCC problem of $Z' \neq Z$ — as shown in paper [3.8]. Physically it is here a consequence of the fact that the TCC problem allows a separation of variables in the elliptic coordinates [3.8]. As for the quasicrossings, they occur when two wells, corresponding to separated $Z$- and $Z'$-centers, have states $\Psi$ and $\Psi'$, characterized by the same energies $E = E'$, by the same magnetic quantum numbers $m = m'$, and by the same radial elliptical quantum numbers $k = k'$ [3.9–3.11]. In this situation, the electron has a much larger probability of tunneling from one well to the other (i.e., of charge exchange) as compared with the absence of such degeneracy.

These rich features of the TCC problem also manifest in a different area of physics such as plasma spectroscopy as follows. A quasicrossing of the TCC terms, by enhancing charge exchange, can result in unusual structures (dips) in the spectral line profile emitted by a $Z$-ion from a plasma consisting of both $Z$- and $Z'$-ions, as was shown theoretically and experimentally [3.12–3.17]. From the experimental shape of these dips it is possible to determine rates of charge exchange between multicharged ions, which is a fundamental reference data virtually inaccessible by other experimental methods [3.17].

Before year 2000, the paradigm was that the above sophisticated features of the TCC problem and its flourishing applications were inherently quantum phenomena. But then in year 2000, papers [3.18, 3.19] were published presenting a purely classical description of both the crossings of energy levels in the TCC problem and the dips in the corresponding spectral line profiles caused by the crossing (via enhanced charge exchange). These classical results were obtained analytically based on first principles without using any model assumptions.

In the classical studies, the TCC systems represent *diatomic* *Rydberg quasimolecules* encountered, e.g., in plasmas containing more than one kind of multicharged ions. Naturally, the classical approach is well-suited for Rydberg quasimolecules.

Later applications of the results from papers [3.18, 3.19] included a magnetic stabilization of Rydberg quasimolecules [3.20], a problem of continuum lowering (CL) in plasmas [3.21] (which plays a key role in calculations of the equation of state, partition function, bound-free opacities, and other collisional atomic transitions in plasmas), and the study of the classical Stark effect for Rydberg quasimolecules (paper [3.22] and Sec. 3.5).

In these studies a particular attention was given to circular Rydberg states (CRS). Circular states of atomic and molecular systems are an important subject in its own right. They have been extensively studied both theoretically and experimentally for several reasons (see, e.g., [3.18–3.20, 3.23–3.37] and references therein): (a) they have long radiative lifetimes and highly anisotropic collision cross-sections, thereby enabling experiments on inhibited spontaneous emission and cold Rydberg gases, (b) these classical states correspond to quantal coherent states, objects of fundamental importance, (c) a classical description of these states is the primary term in the quantal method based on the $1/n$-expansion, and (d) they can be used in developing atom chips.

As examples of experimental studies of Rydberg states, we refer to paper [3.23] where such studies made in the last three decades have been enumerated. In particular, Day and Ebel [3.38] predicted theoretically in 1979 that probability of a wake electron being captured by fast-moving ions traversing a solid to a state with large principal ($n$) and angular momentum ($l$) quantum numbers is quite high and much of the time the electron is captured CRS ($l = n - 1$) distributed over a narrow band near $n_{max}$. Note here that $l = n - 1$ defines circular orbits, whereas the full qualification of CRS requires $|m_l| = l = n - 1$. Day and Ebel proposed existence of an "optical window" in ion velocity as a possible explanation for non-observability of the CRS in beam-foil spectroscopy work. Also, the CRS are both long lived with respect to radiative transitions and short lived with respect to collisions, hence their observation requires a wide aperture and very good vacuum. Pegg *et al.* in 1977 [3.39] observed a strong cascade tails in the decay curves of $Cu^{18+}$ in a beam-foil interaction and attributed it to the successive decay of long-lived CRS or "yrast states". Note that CRS can radiate only to the next lower state, which leads to a chain of successive yrast transitions till they reach to the ground state. Recently, from the study of the time-resolved beam-foil X-ray spectra of projectile or projectile-like ions of 2p, 2s → 1s transitions in H-like Fe, Ni, Cu, and Zn at different delay times (in the range 250–1600 ps). Nandi identified, in each case, a single circular Rydberg and/or an elliptic Rydberg state cascading successively to the 2p or 2s level.

The paradigm was that the above sophisticated features of the TCC problem and its flourishing applications are inherently quantum phenomena. We disprove this paradigm. Here we present a *purely classical description of the crossings of energy terms in the TCC problem leading to charge exchange.* Our classical description is based on first principles and does not use any model assumptions.

In this chapter, we also present detailed studies of modifications of the classical energy terms, their crossings, and charge exchange under various external factors. Those factors, such as a magnetic field, or an electric field, or the screening by plasma electrons, lead to quite different physical outcomes.

There are also two additional classical studies using (as the starting point) the same formalism as above, though not dealing with crossings of energy terms. One of these studies portrays classically muonic-electronic negative hydrogen ion, i.e., the system where an electron and a muon are bound to a proton. It employs also the separation of rapid and slow subsystems. This study is presented in Appendix A.

In another work, the object of a classical study is diatomic Rydberg quasimolecules in a laser field. It reveals laser-field-caused satellites of spectral lines emitted by the quasimolecules. This study is presented in Appendix B.

## 3.2. Helical States of Diatomic Rydberg Quasimolecules

We consider a TCC system, where the charge Z is at the origin and the Oz axis is directed to the charge Z', which is at $z = R$ (here and below the atomic units $\hbar = e = m_e = 1$ are used). In the cylindrical coordinates $(z, \rho, \varphi)$, the Hamiltonian of the system has the form

$$H = (p_z^2 + p_\rho^2 + p_\varphi^2/\rho^2)/2 + U(z, \rho), \qquad (3.1)$$

where the potential energy is

$$U(z, \rho) = -Z/(z^2 + \rho^2)^{1/2} - Z'/[(R - z)^2 + \rho^2]^{1/2}. \qquad (3.2)$$

The relation between the momenta and the corresponding velocities follows from the Hamiltonian equations of the motion:

$$dz/dt = \partial H/\partial p_z = p_z, \quad d\rho/dt = \partial H/\partial p_\rho = p_\rho,$$
$$d\varphi/dt = \partial H/\partial p_\varphi = p_\varphi/\rho^2. \qquad (3.3)$$

Since H does not depend on $\varphi$, the corresponding momentum is conserved:

$$p_\varphi = \rho^2 d\varphi/dt \equiv M = \text{const.} \tag{3.4}$$

Physically, the separation constant $M$ is a projection of the angular momentum on the internuclear axis. Thus, the $z$- and $\rho$-motions can be determined separately from the $\varphi$-motion. Then the $\varphi$-motion can be found from the $\rho$-motion via Eq. (3.4).

The Hamiltonian for the $z$- and $\rho$-motions can be represented in the form

$$H = (p_z^2 + p_\rho^2)/2 + U_{\text{eff}}(z,\rho), \tag{3.5}$$

where an effective potential energy (EPE) is:

$$U_{\text{eff}}(z,\rho) = M^2/(2\rho^2) + U(z,\rho). \tag{3.6}$$

We introduce scaled (dimensionless) variables $w$ and $v$, a scaled projection of the angular momentum $m$, as well as a ratio of the nuclear charges $b$:

$$w \equiv z/R, \quad v \equiv \rho/R, \quad m \equiv M/(ZR)^{1/2}, \quad b \equiv Z'/Z. \tag{3.7}$$

Then the EPE can be re-written as

$$U_{\text{eff}} \equiv (Z/R)u_{\text{eff}}(w,v,m,b),$$
$$u_{\text{eff}}(w,v,m,b) = m^2/(2v^2) - (w^2 + v^2)^{-1/2} - b[(1-w)^2 + v^2]^{-1/2}. \tag{3.8}$$

Now we seek equilibrium points of the EPE. We equate to zero its derivative with respect to $w$

$$\partial U_{\text{eff}}/\partial w = (Z/R)\partial u_{\text{eff}}/\partial w = 0, \tag{3.9}$$

and find a relation

$$b/\left[(1-w)^2 + v^2\right]^{3/2} = w/\left[(1-w)(w^2+v^2)^{3/2}\right], \tag{3.10}$$

which determines a line $v_0(w)$ in the plane $(w,v)$, where the equilibrium points are located:

$$v_0(w,b) = \left\{[w^{2/3}(1-w)^{4/3} - b^{2/3}w^2]/[b^{2/3} - w^{2/3}/(1-w)^{2/3}]\right\}^{1/2}. \tag{3.11}$$

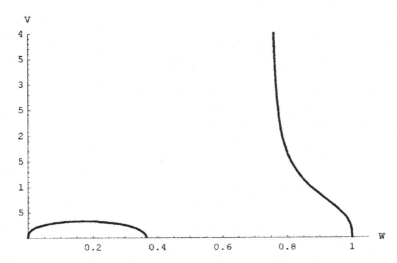

Fig. 3.1. The dependence of the equilibrium value of the scaled radius $v \equiv \rho/R$ of the electron orbit on the scaled axial coordinate $w \equiv z/R$ of the electron for the ratio of nuclear charges $Z'/Z = 3$.

For $b < 1$, the equilibrium value of $v$ exists for $0 \leq w < b/(1 + b)$ and for $1/(1 + b^{1/2}) \leq w \leq 1$. For $b > 1$, the equilibrium value of $v$ exists for $0 \leq w \leq 1/(1 + b^{1/2})$ and for $b/(1 + b) < w \leq 1$. For $b = 1$, the equilibrium value of $v$ exists for the entire range of $0 \leq w \leq 1$. Below we refer to these intervals as the "allowed ranges" of $w$. As an example, Fig. 3.1 shows the dependence $v_0(w)$ for $b = 3$.

By equating to zero the derivative of the EPE with respect to $v$

$$\partial U_{\text{eff}}/\partial v = (Z/R)\partial u_{\text{eff}}/\partial v = 0 \qquad (3.12)$$

and then substituting $v_0(w, b)$ from Eq. (3.10) instead of $v$, we find:

$$m = v_0^2(w, b)/\{(1 - w)^{1/2}[w^2 + v_0^2(w, b)]^{3/4}\} \equiv m_0(w, b). \qquad (3.13)$$

While deriving Eq. (3.13), we used Eq. (3.10) to eliminate an explicit dependence on $b$, so that $b$ enters Eq. (3.13) only implicitly — as an argument of the function $v_0(w, b)$. In a number of subsequent derivations, we will also use Eq. (3.10) for the same purpose without a further notice.

For each set of $(b, m)$, Eq. (3.13) determines one or more values of $w_i(b, m)$ — and then Eq. (3.11) determines the corresponding values of $v_i = v_0(w_i(b, m), b)$ — such that at $(w_i, v_i)$ there is either a stable or unstable equilibrium (the stability of the equilibria will be studied in the next section). A physically equivalent viewpoint, more convenient for the further analysis, is the following: for each set of $(w, b)$, where $w$ belongs to the allowed ranges, Eq. (3.13) determines an equilibrium value of $m_0(w, b)$ — in addition to the equilibrium value of $v_0(w, b)$ determined by Eq. (3.11). Thus, for each b we deal with sets of equilibrium values $(w_i, v_{0i}, m_{0i})$, where $w_i$ belongs to the allowed ranges and $v_{0i} \equiv v_0(w_i)$, $m_{0i} \equiv m_0(w_i)$.

Now, for some value of $b$, we consider a set of equilibrium values $(w_i, v_{0i}, m_{0i})$ and expand the EPE $u_{\text{eff}}$ in terms of $\delta w$ and $\delta v$, where

$$\delta w \equiv w - w_i, \quad \delta v \equiv v - v_{0i}. \tag{3.14}$$

The expansion has the form

$$u_{\text{eff}} \approx u_0 + u_{ww}(\delta w)^2/2 + u_{vv}(\delta v)^2/2 + u_{wv}(\delta w)(\delta v),$$
$$u_0 \equiv u_{\text{eff}}(w_i, v_{0i}, m_{0i}). \tag{3.15}$$

In the subsequent formulas we drop the suffix $i$ for brevity. The second derivatives of the EPE in Eq. (3.15) are

$$u_{ww} \equiv (\partial^2 u_{\text{eff}}/\partial w^2)_0 = [1/(1-w) - 3wP/Q^2]/(w^2 + v_0^2)^{3/2},$$
$$u_{vv} \equiv (\partial^2 u_{\text{eff}}/\partial v^2)_0 = [1/(1-w) + 3wP/Q^2]/(w^2 + v_0^2)^{3/2},$$
$$u_{wv} \equiv (\partial^2 u_{\text{eff}}/\partial w\partial v)_0 = 3v_0 w(2w - 1)/[(w^2 + v_0^2)^{3/2}Q^2], \tag{3.16}$$

where the suffix 0 at the derivatives means that after the differentiation one should set $v = v_0(w)$ and $m = m_0(w)$. In Eq. (3.16) we introduced the following notations:

$$P \equiv w(1 - w) + v_0^2,$$
$$Q \equiv (w^2 + v_0^2)^{1/2}[(1 - w)^2 + v_0^2]^{1/2}. \tag{3.17}$$

Since generally $u_{wv} \neq 0$, a rotation of the reference frame is required in order to transform the EPE to so-called "normal" coordinates, diagonalizing the matrix of the second derivatives of the EPE

[3.40, 3.41]:

$$\delta w' = \delta w \cos \alpha + \delta v \sin \alpha, \quad \delta v' = -\delta w \sin \alpha + \delta v \cos \alpha. \quad (3.18)$$

It is easy to find that

$$\tan 2\alpha = 2u_{wv}/(u_{ww} - u_{vv}) = (1 - 2w)v_0/P, \quad (3.19)$$

so that

$$\cos \alpha = [(1 + P/Q)/2]^{1/2}, \quad \sin \alpha = [(1 - P/Q)/2]^{1/2}\text{sign}(1 - 2w). \quad (3.20)$$

In the normal coordinates, the EPE takes the form

$$u_{\text{eff}} \approx u_0 + \delta w'^2 \omega_-^2 /2 + \delta v'^2 \omega_+^2 /2, \quad (3.21)$$

where

$$\omega_\pm \equiv [1/(1 - w) \pm 3w/Q]^{1/2}/(w^2 + v_0^2)^{3/4}. \quad (3.22)$$

We note that $\omega_+$ is always real. Physically, it is a scaled (dimensionless) frequency of small oscillations around the equilibrium in the direction of the normal coordinate $\delta v'$. In this section, any frequency $F$ and its scaled (dimensionless) counterpart $f$ are related as follows:

$$f \equiv (R^3/Z)^{1/2}F. \quad (3.23)$$

As for the quantity $\omega_-$, it is real if

$$Q \geq 3w(1 - w). \quad (3.24)$$

Using the definition of $Q$ from Eq. (3.17), the condition (3.24) can be re-written as:

$$v_0(w, b) \geq \{w(1 - w) - 1/2 + [9w^2(1 - w)^2 - w(1 - w) + 1/4]^{1/2}\}^{1/2}$$
$$\equiv v_{\text{crit}}(w). \quad (3.25)$$

Physically, under the condition (3.25), the quantity $\omega_-$ is the frequency of small oscillations around the equilibrium in the direction of the normal coordinate $\delta w'$.

Thus, if $v_0(w, b) > v_{\text{crit}}(w)$, the EPE has a two-dimensional minimum at the equilibrium values of $w$ and $v = v_0(w, b)$, so that the equilibrium is stable. Introducing a scaled (dimensionless) time

$$\tau \equiv (Z/R^3)^{1/2}t, \tag{3.26}$$

we obtain the final expression for the small oscillations around the stable equilibrium in the form:

$$\delta w(\tau) = a_w[\cos(\omega_-\tau + \psi_w)]\cos\alpha - a_v[\cos(\omega_+\tau + \psi_v)]\sin\alpha,$$
$$\delta v(\tau) = a_w[\cos(\omega_-\tau + \psi_w)]\sin\alpha + a_v[\cos(\omega_+\tau + \psi_v)]\cos\alpha. \tag{3.27}$$

Here amplitudes $a_w$, $a_v$ and phases $\psi_w$, $\psi_v$ are determined by initial conditions; $\sin\alpha$ and $\cos\alpha$ are given by Eq. (3.20).

Equation (3.4) for the $\varphi$-motion can be re-written in the scaled notations as

$$d\varphi/d\tau = m/v^2. \tag{3.28}$$

Substituting in Eq. (28) $v(\tau) \approx v_0 + \delta v(\tau)$, where $\delta v(\tau)$ is given by Eq. (3.27), and integrating over $\tau$, we obtain the solution for the $\varphi$-motion

$$\varphi(\tau) \approx f\tau - 2f\{\omega_-^{-1}a_w[\sin(\omega_-\tau + \psi_w) - \sin\psi_w]\sin\alpha$$
$$+ \omega_+^{-1}a_v[\sin(\omega_+\tau + \psi_v) - \sin\psi_v)]\cos\alpha\}/v_0, \tag{3.29}$$

where

$$f \equiv (1 - w)^{-1/2}[w^2 + v_0^2]^{-3/4} \tag{3.30}$$

is a scaled (dimensionless) primary frequency of the $\varphi$-motion. Equations (3.28) and (3.29) show that the $\varphi$-motion is a rotation about the internuclear axis with the frequency $f$, slightly modulated by oscillations of the scaled radius of the orbit $v$ at the frequencies $\omega_+$ and $\omega_-$. In other words, the motion of the electron occurs on a *conical surface* of the averaged radius $v_0(w, b)$.

It is interesting to note the following relation between $\omega_\pm$ and $f$ valid *for any b*:

$$[(\omega_+^2 + \omega_-^2)/2]^{1/2} = f. \tag{3.31}$$

If $v_0(w, b) > v_{\text{crit}}(w)$, so that the equilibrium is stable, the relation (3.31) physically means that the rms frequency of the small oscillations of $w$ and $r$ is equal to the primary frequency of the $\varphi$-motion for

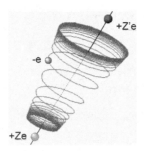

Fig. 3.2. Sketch of the helical motion of the electron in the $ZeZ'$-system at the absence of the magnetic field. We stretched the trajectory along the internuclear axis to make its details better visible.

any $b$. If $v_0(w, b) < v_{crit}(w)$, the equilibrium is unstable: the quantity $\omega_-$ takes imaginary values; its absolute value $|\omega_-|$ is the increment of the instability developing in the direction of the normal coordinate $\delta w'$. However, even for the unstable equilibrium, the relation (3.31) still holds for any $b$.

Thus, for the stable motion, the electron trajectory is a helix on the surface of a cone, with the axis coinciding with the internuclear axis. In this *helical* state, the electron, while spiraling on the surface of the cone, oscillates between two end-circles which result from cutting the cone by two parallel planes perpendicular to its axis (Fig. 3.2).

## 3.3. Crossings of Classical Energy Terms of Diatomic Rydberg Quasimolecules

In this section, our goal is to find out whether or not classical mechanics reproduces crossings of energy terms of the *same symmetry*. In the quantum TCC problem, "terms of the same symmetry" means terms of the same magnetic quantum number $m$ [3.7–3.11]. Therefore, in our classical TCC problem, from now on we fix the angular momentum projection $M$ and study the behavior of the classical energy at $M = \text{const} \geq 0$ (the results for $M$ and $-M$ are physically the same). Since $m \equiv M/(ZR)^{1/2}$ in accordance to Eq. (7), then using Eq. (3.13) we obtain

$$R(w, b, M) = M^2/[Zm_0^2(w, b)], \qquad (3.32)$$

so that $z(w, b, M) = wR(w, b, M)$ and $\rho(w, b, M) = v_0(w, b)R(w, b, M)$. For any $b > 0$, for any $w$ from the allowed ranges (controlled by the value of $b$), and for any $M \geq 0$, the internuclear distance $R$, the location of the orbital plane $z$, and the radius of the orbit $\rho$ have each its individual unique equilibrium value given by the functions $R(w, b, M)$, $z(w, b, M)$, and $\rho(w, b, M)$, respectively.

For simplicity, we consider the limit where $a_w = a_v = 0$. In other words, here we disregard the small oscillations of the $z$- and $\rho$-coordinates and focus at the primary motion (i.e., at the $\varphi$-motion), where the electron moves around the circle of the radius $\rho(w, b, M) = v_0(w, b)R(w, b, M)$. In this situation, the total energy $E$ of the electron coincides with the EPE given by Eq. (3.6), so that

$$E = M^2/(2\rho^2) + U(z, \rho). \tag{3.33}$$

Similarly to Eq. (3.8), we introduce a scaled total energy $e$ as follows:

$$E \equiv (Z/R)e[w, v_0(w, b), m_0(w, b), b],$$
$$e[w, v_0(w, b), m_0(w, b), b]$$
$$= m_0^2/(2v_0^2) - (w^2 + v_0^2)^{-1/2} - b[(1 - w)^2 + v_0^2]^{-1/2}. \tag{3.34}$$

Using Eqs. (3.10) and (3.13), the latter quantity can be re-written as

$$e(w, b) = -[w(1 - w) + v_0^2(w, b)/2]/\{(1 - w)[w^2 + v_0^2(w, b)]^{3/2}\}. \tag{3.35}$$

Now we substitute $R(w, b, M)$ from Eq. (3.32) in Eq. (3.34) and find:

$$E(w, b, M) = (Z/M)^2 m_0^2(w, b)e(w, b). \tag{3.36}$$

Thus, for any $b > 0$ and $M \geq 0$, Eqs. (3.32) and (3.35) determine in a parametric form (via $w$) the dependence of the energy $E$ on the internuclear distance $R$, i.e., the *classical energy terms*.

Figure 3.3 shows the dependence of the scaled energy $(M/Z)^2E$ on the scaled internuclear distance $(Z/M^2)R$ (both quantities are dimensionless) for $b = 3$. The results are astonishing. There is more than one classical energy term — namely, there are three terms of the same symmetry. Just this is already counterintuitive. Moreover,

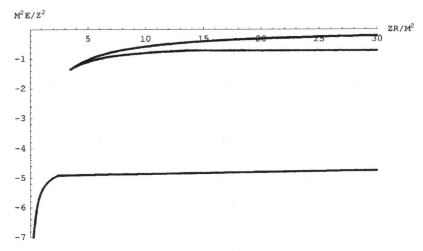

Fig. 3.3. Classical energy terms: The dependence of the scaled classical energy $(M/Z)^2 E$ on the scaled internuclear distance $(Z/M^2)R$ for $Z' = 3Z$.

two of these classical energy terms cross. (We call it V-type crossing since it resembles the inclined letter V.)

We emphasize that the above example of $Z'/Z = 3$ represents a typical situation. In fact, *for any pair of Z and Z'* $\neq Z$ there are three classical energy terms of the same symmetry and *the upper term always crosses the middle term*. (For $Z' = Z$ there is only one term in the corresponding plot and no crossing — as should be expected.)

For arbitrary $Z'/Z \neq 1$, these three terms have the following origin. At $R \to \infty$, the lower term corresponds to the energy of the hydrogen-like ion (HI) of the nuclear charge $Z_{max} \equiv \max(Z', Z)$, $(E \to -(Z_{max}/M)^2/2)$, slightly perturbed by the charge $Z_{min} \equiv \min(Z', Z)$. This ion is in a Stark state, corresponding classically to the zero projection $A_0$ of the Runge–Lenz vector [3.40] on the axis Oz and corresponding quantum-mechanically to the zero electric quantum number $q \equiv n_1 - n_2$ ($n_1$, $n_2$ are the parabolic quantum numbers [3.42]) — as should be expected for the circular states. At $R \to 0$, the lower term corresponds to the energy of the HI of the nuclear charge $Z' + Z$, $(E \to -[(Z + Z')/M]^2/2)$, i.e., to the so-called "united atom" [3.7–3.11].

At $R \to \infty$, the middle term corresponds to the energy of the HI of the nuclear charge $Z_{min}$ $(E \to -(Z_{min}/M)^2/2)$, slightly perturbed by the charge $Z_{max}$. This ion is in a Stark state, corresponding classically to $A_0 = 0$ and corresponding quantum-mechanically to $q = 0$ — as should be expected for the circular states.

At $R \to \infty$, the upper term corresponds to a near-zero-energy state (where the electron is almost free). In terms of the parameter $w$, this term at $R \to \infty$ corresponds to $w \to 1/(1 + b^{1/2}) \equiv w_0$. If the ratio $Z'/Z$ is of the order of (but not equal to) unity, the upper term at $R \to \infty$ can be described only in the terminology of elliptical coordinates (rather than parabolic or spherical coordinates), meaning that even at $R \to \infty$ the electron is shared between the $Z$- and $Z'$-centers. However, in the case of $Z' \gg Z$, the potential at $w = w_0$ is very close to the potential due to an isolated charge $Z'$, so that the upper term can be asymptotically considered as the $Z'$-term. It crosses the middle term, which asymptotically is the $Z$-term (since $Z_{min} = Z$ for $Z' > Z$). Likewise, in the case of $Z' \ll Z$, the potential at $w = w_0$ is very close to the potential due to an isolated charge $Z$, so that the upper term can be asymptotically considered as the Z-term. It crosses the middle term, which asymptotically is the $Z'$-term (since $Z_{min} = Z'$ for $Z' < Z$).

Thus, when Z and Z' differ significantly from each other, the crossing occurs between two classical energy terms which can be asymptotically labeled as $Z$- and $Z'$-terms. This situation *classically depicts charge exchange*. Indeed, say, initially at $R \to \infty$, the electron was a part of the HI of the nuclear charge $Z_{min}$. As the charges Z and $Z'$ come relatively close to each other, the middle and upper classical terms cross and the electron is shared between the $Z$- and $Z'$-centers. Finally, as the charges Z and $Z'$ go away from each other, the electron ends up as a part of the HI of the nuclear charge $Z_{max}$.

Physically, crossings and quasicrossings in the TCC problem are possible only because the problem possesses an additional (to $E$ and $M$) conserved quantity (integral of motion) [3.43]

$$A = (1/2)(\mathbf{p} \times \mathbf{M} - \mathbf{M} \times \mathbf{p})$$
$$\cdot \mathbf{e_z} - M^2/R - Zz/r - Z'(R - z)/|\mathbf{R} - \mathbf{r}| + Z', \quad \mathbf{e_z} = \mathbf{R}/R,$$
$$(3.37)$$

where **p** is a linear momentum vector, the suffix $z$ means a $z$-projection. The last term in Eq. (3.37) was added to facilitate the transition at $R \to \infty$ to the Stark effect for the HI of the nuclear charge $Z$ slightly perturbed by the charge $Z'$. The quantity $A$ is a $z$-projection of a generalized Runge–Lenz vector [3.43]. For our classical circular TCC states this quantity becomes:

$$A(w, b) = Za(w, b),$$
$$a(w, b) \equiv w[w^2 + (1 - w)^2 + v_0^2(w, b)(1 - 2w)/(1 - w)]/$$
$$[w^2 + v_0^2(w, b)]^{3/2} - m_0^2(w, b) - b. \qquad (3.38)$$

In the quantum TCC problem, two terms of the same symmetry $(m = m')$ should differ by their angular elliptic quantum numbers for a quasicrossing ($s = s' + 1$, while $k = k'$) or by both elliptic quantum numbers for a crossing ($s \neq s'$, $k \neq k'$) [3.9–3.11]. So, in either case it should be $\Delta s \equiv s' - s \neq 0$. To find a classical correspondence to these "selection rules", we recall that for each quantum term, the eigenvalues $E^*$, $A^*$, $M^*$ of the operators $E^\wedge$, $A^\wedge$, $M^\wedge$ are some functions of the quantum numbers $k$, $s$, $m$ (in particular, $M^* = m$). Therefore, the quantum number $s$ should also be some function of $E^*$, $A^*$, $M^*$: $s = g(E^*, A^*, M^*)$. The eigenvalues $E^*$, $A^*$, $M^*$ correspond to the classical quantities $E(w)$, $A(w)$, $M(w)$ of our circular TCC states, so that $s = g[E(w), A(w), M(w)]$. It turns out that at the value of $w = w_c$, corresponding to the crossing of the upper and middle classical terms, we have:

$$E'(w_c) = M'(w_c) = 0, \quad A'(w_c) \neq 0, \qquad (3.39)$$

where the sign' stands for the derivative with respect to $w$. Consequently, in the vicinity of $w_c$ we have: $\Delta s = [(\partial g/\partial E)E'(w_c) + (\partial g/\partial A)A'(w_c) + (\partial g/\partial M)M'(w_c)]\Delta w = (\partial g/\partial A)A'(w_c)\Delta w$. So, if it were not for the presence of $A$ and for the fact that $A'(w_c) \neq 0$, we would have had $\Delta s = 0$ — contrary to the above selection rules. Thus, the classical crossings are intimately connected with the dynamical symmetry of the TCC problem, and the remarkable set of "rules" from Eq. (3.39) is a classical counterpart of the above quantum selection rules.

We also note that at the crossing we have $\omega_- = 0$, where $\omega_-$ is given by Eq. (3.22). So, among the two crossing classical energy terms, the middle term corresponds to stable equilibria of the three-dimensional motion, while the upper term corresponds to unstable equilibria of the three-dimensional motion. Thus, the value of $w_c$ corresponding to the crossing can be determined as a root of either one of the following three equations: from $E'(w_c) = 0$, or from $M'(w_c) = 0$, or from

$$v_0(w_c, b) = v_{\text{crit}}(w_c), \qquad (3.40)$$

where $v_{\text{crit}}(w)$ is defined in Eq. (3.25). In any case, one should accept only the root $w_c$ within the interval $(0,1)$, the endpoints 0 and 1 being excluded.

It turns out that the form of the parametric dependence can be significantly simplified and some of the properties of it can be found analytically by introducing a new parameter

$$\gamma = \left( \frac{1}{w} - 1 \right)^{1/3}. \qquad (3.41)$$

In this case, $w = 0$ will correspond to $\gamma = +\infty$ and $w = 1$ will correspond to $\gamma = 0$, thus $\gamma > 0$ in the allowed regions. The points $w_1 = 1/(1 + b^{1/2})$ and $w_3 = b/(1 + b)$ defining the allowed regions $0 < w < w_1$, $w_3 < w < 1$ (here we assume $b > 1$) will correspond to $\gamma_1 = b^{1/6}$ and $\gamma_3 = 1/b^{1/3}$ (notice that $0 < w < w_1$ corresponds to $+\infty > \gamma > \gamma_1$ and $w_3 < w < 1$ corresponds to $\gamma_3 > \gamma > 0$).

From now on we will denote the scaled internuclear distance $(Z/M^2)R$ as $r$:

$$(Z/M^2)R = r = 1/m_0^2, \qquad (3.42)$$

where the second equality in Eq. (3.42) is the consequence of Eq. (3.32). Also from now on we will use for the scaled energy $(M/Z)^2 E$ the notation $-\varepsilon_1$ and for the scaled Hamilton function (called in this book classical Hamiltonian) the notation $h$:

$$(M/Z)^2 E = -\varepsilon_1, \quad (M/Z)^2 H = h. \qquad (3.43)$$

Obviously, $h = -\varepsilon_1$, since classically $H = E$.

The scaled energy terms $-\varepsilon_1(r)$ will take the following form in the $\gamma$-parametrization defined by Eq. (3.41):

$$-\varepsilon_1(\gamma, b) = -\frac{(b^{2/3} - \gamma^4)^2(\gamma(\gamma^3 - 2) + b^{2/3}(2\gamma^3 - 1))}{2(\gamma^3 - 1)^2(\gamma^6 - 1)}, \quad (3.44)$$

$$r(\gamma, b) = \frac{\sqrt{b^{2/3}\gamma^2 - 1}(\gamma^6 - 1)^{3/2}}{\gamma(b^{2/3} - \gamma^4)^2}. \quad (3.45)$$

The parametric plot of $-\varepsilon_1(\gamma, b)$ from Eq. (3.44) versus $r(\gamma, b)$ from (3.45) with the parameter $\gamma$ varied from 0 to $1/b^{1/3}$ and from $b^{1/6}$ to $+\infty$ for $b = 3$ will yield the same graph as in Fig. 3.3.

## 3.4. Effects of a Static Magnetic Field: Stabilization of Diatomic Rydberg Quasimolecules

We consider here CRS of the $ZeZ'$-system in the presence of a uniform magnetic field $B$ parallel to the internuclear axis. For $z \ll R$ or for $(R-z) \ll R$ when the electron is mainly bound to the $Z$ or the $Z'$ ion and is perturbed by the other fully stripped ion, these circular orbits depict Stark states which correspond classically to zero projection of the Runge–Lenz vector [3.40] on the axis OZ and quantally to zero electric quantum number $k = n_1 - n_2$, where $n_1$ and $n_2$ are the parabolic quantum numbers [3.42]. The classical Hamiltonian is given by (in atomic units):

$$\begin{aligned}
H(\rho, z) &= M^2/(2\rho^2) - Z/(\rho^2 + z^2)^{1/2} - Z'/[\rho^2 + (z - R)^2]^{1/2} \\
&\quad + \Omega M + \Omega^2\rho^2/2, \quad \Omega \equiv B/(2c). \quad (3.46)
\end{aligned}$$

Here $M$ is the constant $z$-component of the angular momentum and $\Omega$ is the Larmor frequency expressed in practical units as $\Omega(s^{-1}) \approx 8.794 \times 10^6$ B(G).

Introduce the following scaled quantities:

$$\begin{aligned}
b &\equiv Z'/Z, \quad u \equiv \rho/R, \quad w \equiv z/R, \quad m \equiv M/(ZR)^{1/2}, \\
\omega &\equiv \Omega M^3/Z^2, \quad h \equiv HM^2/Z^2 = M^2/Z^2, \quad (3.47)
\end{aligned}$$

where $E$ is the energy. In these notations, the scaled Hamiltonian/energy is

$$h(u, w, \omega) = m^2 \eta(u, w, \omega),$$
$$\eta(u, w, \omega) \equiv m^2/(2u^2) - 1/(u^2 + w^2)^{1/2} - b/[u^2 + (1 - w)^2]^{1/2}$$
$$+ \omega/m^2 + \omega^2 u^2/(2m^6). \qquad (3.48)$$

The conditions for dynamic equilibrium are

$$\partial h/\partial w = m^2\{w/(u^2+w^2)^{3/2} - b(1-w)/[u^2+(1-w)^2]^{3/2}\} = 0 \quad (3.49)$$

and

$$\partial h/\partial u = m^2\{-m^2/u^3 + u/(u^2 + w^2)^{3/2} + bu/[u^2 + (1 - w)^2]^{3/2}$$
$$+ \omega^2 u/m^6\} = 0. \qquad (3.50)$$

Equation (3.49) shows that equilibrium along the internuclear axis does not depend on the scaled magnetic field $\omega$. In terms of the equilibrium value $w_0$ of $w$, the equilibrium value of $u$ can therefore be expressed as

$$u(w_0, b) = \{[w_0(1-w_0)^2]^{2/3} - b^{2/3}w_0^2\}^{1/2}/\{b^{2/3} - [w_0/(1-w_0)]^{2/3}\}^{1/2}, \qquad (3.51)$$

which only exists within the following "allowed ranges",

$$0 \le w_0 < b/(1+b) \quad \text{and} \quad 1/(1+b^{1/2}) \le w_0 \le 1; \quad b < 1;$$
$$0 \le w_0 \le 1/(1+b^{1/2}) \quad \text{and} \quad b/(1+b) < w_0 \le 1; \quad b > 1; \qquad (3.52)$$
$$0 \le w_0 \le 1; \qquad\qquad\qquad\qquad\qquad\qquad\qquad b = 1$$

of $w_0$. Equation (3.50), represents the condition for equilibrium in the orbital plane and can be re-written in the form

$$m(w_0, b, \omega) = \pm\{f/4 + (f^2/4 + j)^{1/2}/2$$
$$+ [f^2/2 - j + (f^3/4)/(f^2/4 + j)^{1/2}]^{1/2}/2\}^{1/2}, \qquad (3.53)$$

where, in terms of $u(w_0, b)$, given by Eq. (3.51),

$$f(w_0, b, \omega) \equiv u^4(w_0, b)/[u^2(w_0, b) + w_0^2]^{3/2}$$
$$+ bu^4(w_0, b)/[u^2(w_0, b) + (1 - w_0)^2]^{3/2}, \quad (3.54)$$

and

$$j(w_0, b, \omega) \equiv [u^4(w_0, b)\omega^2/18]^{1/3}g - (4/g)[2u^8(w_0, b)\omega^4/3]^{1/3},$$

(3.55)

with

$$g(w_0, b, \omega) \equiv |-9f^2 + [81f^4 + 768u^4(w_0, b)\omega^2]^{1/2}|^{1/3}.$$ 
(3.56)

The plus and minus signs in Eq. (3.53) correspond, respectively to the positive and negative projections of the angular momentum along the magnetic field. For each set $\{b, m, \omega\}$ of parameters, Eq. (3.53) determines the equilibrium value $w_0$ of the scaled $z$-coordinate of the orbital plane.

The internuclear distance $R$ was considered to be "frozen". In order to reproduce the electronic terms, i.e., the dependence of the electronic energy on the internuclear distance, one should now allow $R$ to be a slowly varying adiabatic quantity (slowly varying with respect to the electronic motion, as in the Born–Oppenheimer approach [3.44]).

We consider energy terms of the *same symmetry* which, for the quantal $ZeZ'$-problem, means terms with the same magnetic quantum number M [3.7–3.11]. Therefore, in our classical $ZeZ'$-problem, from now on we consider fixed projection of the angular momentum $M$ and study the behavior of the classical energy keeping $M$ constant.

We introduce the scaled internuclear distance similarly to Eq. (3.43):

$$r(w_0, b, \omega) = 1/m^2(w_0, b, \omega).$$

(3.57)

On substituting $w = w_0$ into Eq. (3.48), then,

$$h(w_0, b, \omega) = m^2(w_0, b, \omega)\eta[u(w_0, b), w_0, \omega].$$

(3.58)

Thus, for any positive ratio of the nuclear charges $b > 0$ and for any value of the scaled magnetic field $\omega$, the dependence of the scaled energy $h$ on the scaled internuclear distance $r$ is determined by Eqs. (3.57) and (3.58) in terms of the parameter $w_0$, which takes all values from the allowed ranges specified by Eq. (3.52). In other words,

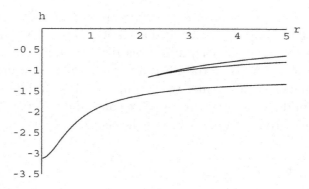

Fig. 3.4. The scaled electronic energy $h$ versus the scaled internuclear distance $r$ for the ratio of the nuclear charges $b = 3/2$ at the absence of the magnetic field ($h$ and $r$ are defined by Eqs. (3.58) and (3.57), respectively).

Eqs. (3.57) and (3.58) determine the *classical electronic energy terms* for any strength of the magnetic field, *including the strong field case.*

Figure 3.4 shows the scaled electronic energy $h$ versus the scaled internuclear distance $r$ for $b = 3/2$ in the absence of magnetic field.

We note that the upper and middle energy terms terminate at some $r = r_{\min}$, so that there are no CRS at $r < r_{\min}$ for these two energy terms. The classical energy of the CRS acquires an imaginary part at $r < r_{\min}$, corresponding quantally to virtual states/resonances. There may well be non-CRS at $r < r_{\min}$ in the same energy range.

At this point it might be useful to clarify the relation between the classical energy terms $h(r)$ and the energy $E$. The former is a scaled quantity related to the energy as specified above: $E = -(Z/M)^2 h$. The projection $M$ of the angular momentum on the internuclear axis is a *continuous* variable. The energy $E$ depends on both $h$ and $M$. Therefore, while the scaled quantity $h$ takes a *discrete* set of values, the energy $E$ takes a *continuous* set of values (as it should be in classical physics).

We now "turn on" the magnetic field. Figure 3.5 shows the scaled electronic energy $h$ versus the scaled internuclear distance $r$ for $b = 3/2$ at $\omega = +1.1$, i.e., at a moderate value of the magnetic field. We note that $\omega > 0$ corresponds to $BM > 0$, while $\omega < 0$ corresponds to

$BM < 0$; remember $B$ and $M$ are the $z$-projections of the magnetic field and of the angular momentum, respectively, and that the Oz axis is directed from the charge $Z$ toward the charge $Z'$.

Figure 3.5 shows that the magnetic field corresponding to $\omega = +1.1$ and higher values, under the condition $BM > 0$, lifts the entire upper and middle energy terms into the continuum. Figure 3.6 shows the scaled electronic energy $h$ versus the scaled internuclear distance

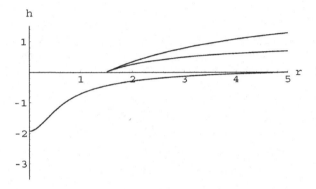

Fig. 3.5. Same as in Fig. 3.4, but at the scaled magnetic field $\omega = +1.1$. We note that $\omega > 0$ corresponds to $BM > 0$, while $\omega < 0$ corresponds to $BM < 0$; here $B$ and $M$ are $z$-projections of the magnetic field and of the angular momentum, respectively; the Oz axis is directed from the charge $Z$ toward the charge $Z'$.

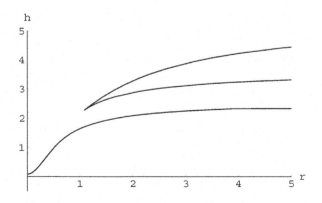

Fig. 3.6. Same as in Fig. 3.5, but at $\omega = +2.7$.

$r$ for $b = 3/2$ at $\omega = +2.7$, i.e., at a larger value of the magnetic field. It is seen that the magnetic field of this value (and of higher values), under the condition $BM > 0$, lifts all three energy terms into the continuum.

These CRS above the ionization threshold, shown in Figs. 3.5 and 3.6, are *classical molecular counterparts of the quantal atomic quasi-Landau levels* or resonances. The latter were discovered experimentally by Garton and Tomkins [3.45] (for theoretical references on atomic quasi-Landau resonances, see, e.g., the book [3.46]).

Now we explore the *stability of the nuclear motion* in the $ZeZ'$-system. The electronic energy $E(R, B)$ becomes a crucial part of the effective internuclear potential

$$V(R, B) = ZZ'/R + E(R, B) \qquad (3.59)$$

for the relative motion of the nuclei. The scaled internuclear potential

$$v \equiv VM^2/Z^2, \qquad (3.60)$$

then reduces (c.f., Eq. (3.58)) to

$$v(w_0, b, Z', \omega) = m^2(w_0, b, \omega)\{\varepsilon[u(w_0, b), w_0, \omega] + Z'\}. \qquad (3.61)$$

For any set $\{b, Z', \omega\}$, Eqs. (3.57) and (3.61) therefore determine the dependence of the scaled internuclear potential $v$ on the scaled internuclear distance $r$ in terms of the parameter $w_0$ which takes all values within the allowed ranges specified by Eq. (3.52). In other words, Eqs. (3.57) and (3.61) determine the *classical effective internuclear potential* for any strength of the magnetic field, including the strong field case.

Figure 3.7 shows the upper and middle branches of the scaled effective internuclear potential $v$ versus the scaled internuclear distance $r$ for $Z = 2$ and $Z' = 3$ in the absence of the magnetic field. It is seen for any starting point at the middle branch, that the system would "find" the way to lowering its potential energy without any obstacle and would end up at an infinitely large internuclear distance, thereby resulting in dissociation.

The same is true for the lower branch (not shown in Fig. 3.7). In other words, in the absence of the magnetic field, the CRS-system,

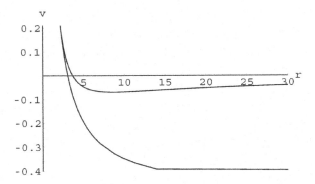

Fig. 3.7. The upper and middle branches of the scaled effective internuclear potential $v$ (defined by Eqs. (3.60) and (3.61)) versus the scaled internuclear distance $r$ for $Z = 2$ and $Z' = 3$ at the absence of the magnetic field.

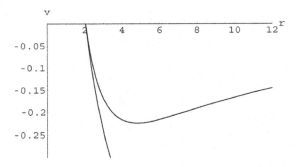

Fig. 3.8. The same as in Fig. 3.7, but at the scaled magnetic field $\omega = -0.3$ (note that $BM < 0$).

associated with the middle or lower branches of the EPE, is not really a molecule, but only a quasimolecule because the molecular orbital is *antibonding*. The corresponding classical result was obtained previously by Pauli [3.47] for the molecular hydrogen ion $H_2^+$. The upper branch in Fig. 3.4 displays a very shallow minimum of $v = -0.0688$ located at $r = 8.7$.

We now "turn on" the magnetic field. Figure 3.8 shows the upper and middle branches of the scaled effective internuclear potential $v$ versus the scaled internuclear distance $r$ for $Z = 2$ and $Z' = 3$ at a relatively small scaled magnetic field $\omega = -0.3$ (with $BM < 0$).

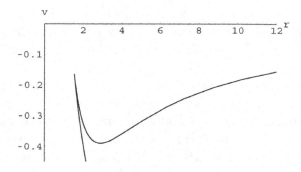

Fig. 3.9. The same as in Fig. 3.8, but at the scaled magnetic field $\omega = -1$.

It is seen that the minimum in the upper branch became significantly deeper and moved towards lower $r$.

We note that the scaled magnetic field $|\omega| = 0.3$ would correspond to the magnetic field $B \sim 10^5$ G for $|M| \sim 30$. The magnetic field $B \sim 10^5$ G would be typical for magnetic fusion devices under construction.

Figure 3.9 shows the same as Fig. 3.8, but for $\omega = -1$. As the magnetic field increased, it is seen that the minimum in the upper branch becomes further deepened and moves even closer to the origin.

The "cusp" formed by the upper and middle branches in Figs. 3.7–3.9 reflects the fact that the upper and middle energy terms for the corresponding electronic terms terminate at some $r = r_{\min}$ — as already noted above. Although present in CRS, this cusp may not appear in non-CRS.

Figures 3.7–3.9 reveal *magnetic stabilization of the nuclear motion* for the case of $BM < 0$. Indeed, in the absence of the magnetic field, the potential well is very shallow. It is known that too shallow potential wells do not have any quantal discrete energy levels (see, e.g., book [3.42]). Moreover, if this system is embedded in a plasma, then due to the known phenomenon of the "continuum lowering" by the plasma environment (see, e.g., books/reviews [3.48–3.50] and references therein, as well as Sec. 3.7), the minimum of this very shallow potential well in Fig. 3.7 could be "absorbed" by the lowered continuum. The magnetic field dramatically deepens the potential well and therefore stabilizes the system for the case of $BM < 0$. The magnetic field can therefore transform the quasimolecule into a real, classically described molecule so that the molecular orbital becomes *bonding*.

The particular example of the system chosen for Figs. 3.7–3.9 corresponds to the CRS of an electron in the vicinity of the nuclei of He and Li. Both nuclei are usually present in magnetic fusion plasmas. Moreover, in such plasmas, Rydberg states of either of these nuclei result from charge exchange with ions of higher nuclear charge that are always present in magnetic fusion plasmas. Relatively large magnetic-field strengths are also present. It should be therefore possible to observe magnetic stabilization of the quasimolecule HeLi$^{4+}$ present in these practically important experimental devices.

Our analysis has also shown that a similar magnetic stabilization of Rydberg quasimolecules in CRS is displayed by other (though not all) $ZeZ'$-systems characterized by the ratio of the nuclear charges in the range: $1 < Z'/Z < 3$. Our results open up this phenomenon for possible further theoretical and experimental investigation.

## 3.5. Effects of a Static Electric Field on Diatomic Rydberg Quasimolecules: Enhancement of Charge Exchange and of Ionization

We consider a TCC system, where the charge $Z$ is at the origin and the $Oz$ axis is directed to the charge $Z'$, which is at $z = R$. A uniform electric field $\mathbf{F}$ is applied along the internuclear axis — in the negative direction of $Oz$ axis. We study CRS where the electron moves around a circle in the plane perpendicular to the internuclear axis, the circle being centered at this axis.

Two quantities, the energy $E$ and the projection $M$ of the angular momentum on the internuclear axis are conserved in this configuration. We use cylindrical coordinates to write the equations for both.

$$E = \frac{1}{2}\left(\dot{\rho}^2 + \rho^2\dot{\varphi}^2 + \dot{z}^2\right) - \frac{Z}{r} - \frac{Z'}{r'} - Fz, \qquad (3.62)$$

$$M = \rho^2\dot{\varphi}, \qquad (3.63)$$

where $\rho$ is the distance of the electron from the internuclear axis, $\varphi$ is its azimuthal angle, $z$ is the projection of the radius-vector of the electron on the internuclear axis, $r$ and $r'$ are the distances of the electron from the particle to $Z$ and $Z'$, respectively.

The circular motion implies that $d\rho/dt = 0$; as the motion occurs in the plane perpendicular to the $z$-axis, $dz/dt = 0$. Further, expressing $r$ and $r'$ through $\rho$ and $z$, and taking $d\varphi/dt$ from Eq. (3.63), we have:

$$E = \frac{M^2}{2\rho^2} - \frac{Z}{\sqrt{\rho^2 + z^2}} - \frac{Z'}{\sqrt{\rho^2 + (R - z)^2}} - Fz. \qquad (3.64)$$

With the scaled quantities

$$w = \frac{z}{R}, \quad v = \frac{\rho}{R}, \quad b = \frac{Z'}{Z}, \quad \varepsilon = -\frac{ER}{Z},$$

$$l = \frac{M}{\sqrt{ZR}}, \quad f = \frac{FR^2}{Z}, \quad r = \frac{ZR}{M^2} \qquad (3.65)$$

our energy equation takes the form below:

$$\varepsilon = \frac{1}{\sqrt{w^2 + v^2}} + \frac{b}{\sqrt{(1 - w)^2 + v^2}} + fw - \frac{l^2}{2v^2}. \qquad (3.66)$$

We seek the equilibrium points by finding partial derivatives of $\varepsilon$ by the scaled coordinates $w, v$ and setting them equal to zero. This yields the following two equations.

$$f + \frac{b(1 - w)}{((1 - w)^2 + v^2)^{3/2}} = \frac{w}{(w^2 + v^2)^{3/2}}, \qquad (3.67)$$

$$l^2 = v^4 \left( \frac{1}{(w^2 + v^2)^{3/2}} + \frac{b}{((1 - w)^2 + v^2)^{3/2}} \right). \qquad (3.68)$$

From the definitions of the scaled quantities (3.65), $\ell^2 = 1/r$ and $E = -(Z/R)\varepsilon$. Since $r = ZR/M^2$, then $E = -(Z/M)^2\varepsilon/r$, where $r = 1/\ell^2$ can be obtained by solving Eq. (3.68) for $\ell$. Substituting $\ell$ into the energy equation, we get the three master equations for this configuration.

$$\varepsilon_1 = p^2 \left( \frac{1}{(w^2 + p)^{3/2}} + \frac{b}{((1 - w)^2 + p)^{3/2}} \right)$$

$$\times \left( \frac{w^2 + p/2}{(w^2 + p)^{3/2}} + \frac{b((1 - w)^2 + p/2)}{((1 - w)^2 + p)^{3/2}} + fw \right), \quad (3.69)$$

$$r = p^{-2} \left( \frac{1}{(w^2 + p)^{3/2}} + \frac{b}{((1 - w)^2 + p)^{3/2}} \right)^{-1}, \quad (3.70)$$

$$f + \frac{b(1 - w)}{((1 - w)^2 + p)^{3/2}} = \frac{w}{(w^2 + p)^{3/2}}, \quad (3.71)$$

where $E = -(Z/M)^2 \varepsilon_1$ and $p = v^2$. Thus, $\varepsilon_1$ is the "true" scaled energy, whose equation for $E$ does not include the internuclear distance $R$. The scaled energy $\varepsilon_1$ and the scaled internuclear distance $r$ in Eqs. (3.69) and (3.70) now explicitly depend only on the coordinates $w$ and $p$ (besides the constants $b$ and $f$). Therefore, if we solve Eq. (3.71) for $p$ and substitute it into Eqs. (3.69) and (3.70), we will have the parametric solution $\varepsilon_1(r)$ with the parameter $w$.

Our focus is at crossings of energy terms of the *same symmetry*, i.e., of the same angular momentum projection $M$. Therefore, we fix $M$ and study the behavior of the classical energy terms at $M = \text{const} \geq 0$ (the results for $M$ and $-M$ are physically the same).

Equation (3.71) does not allow an exact analytical solution for $p$. Therefore, we will use an approximate analytical method.

Figure 3.10 shows a contour plot of Eq. (3.71) for a relatively weak field $f = 0.3$ at $b = 3$, with $w$ on the horizontal axis and $p$ on the vertical. The plot has two branches. The left branch spans from $w = 0$ to $w = w_1$. The right one actually has a small two-valued region between some $w = w_3$ and 1 ($w_3 < 1$). Indeed, at $w = 1$, there are two values of $p$: $p = 0$ and $p = f^{-2/3} - 1$. Thus, the two-valued region exists only for $f < 1$.

The right branch touches the abscissa at $w = 1$ and at some $w = w_2$. Analytical expressions for $w_1$ and $w_2$ are bulky and we do not reproduce them here. They can be found in our paper [3.24].

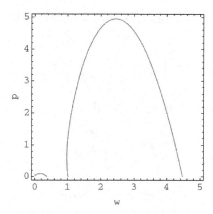

Fig. 3.10. Contour plot of Eq. (3.70) for a relatively weak field $f = FR^2/Z = 0.3$ at $b = Z'/Z = 3$.

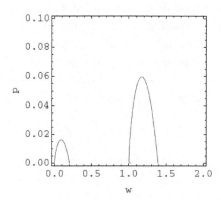

Fig. 3.11. The same as in Fig. 3.10, but for a relatively strong field $f = FR^2/Z = 20$.

As shown in paper [3.24], the quantity $w_3$ is a solution of the equation

$$f^{2/5}(2w_3 - 1)^{3/5} = w_3^{2/5} - b^{2/5}(1 - w_3)^{2/5} \qquad (3.72)$$

Figure 3.11 shows a contour plot of Eq. (3.71) for a relatively strong field $f = 20$ at $b = 3$. It is seen that there is no two-valued region.

From now on we consider the situation where the radius of the electronic orbit is relatively small, meaning that $p \ll 1$. Physically this corresponds to strong fields $f > f_{\min} \sim 10$.

Solving (3.71) in the small-$p$ approximation, we obtain

$$p = \left( \frac{w}{f + \frac{b}{(1-w)^2}} \right)^{2/3} - w^2 \tag{3.73}$$

for the left branch $(0 < w < w_1)$ and

$$p = \left( \frac{b(1-w)}{\frac{1}{w^2} - f} \right)^{2/3} - (1-w)^2 \tag{3.74}$$

for the right branch $(1 < w < w_2)$. Substituting these results into (3.69) and (3.70), we get approximate solutions for energy terms $-\varepsilon_1(r)$ in both regions in a parametric form, $w$ being the parameter.

Now we plot classical energy terms $-\varepsilon_1(r)$ by varying the parameter $w$ over both regions, using the appropriate formula for each one. Figure 3.12 presents classical energy terms at $b = 3$ for $f = 5$.

We also solved the same problem numerically. By comparison we found that the approximate analytical solution is accurate for fields $f = 5$ and above. Figures 3.13 and 3.14 show the numerically obtained classical energy terms at $b = 3$ for $f = 2$ and $f = 0.1$, respectively.

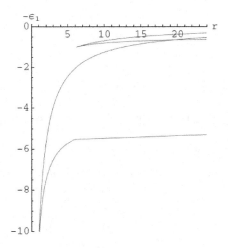

Fig. 3.12. Classical energy terms at $b = 3$ for $f = 5$.

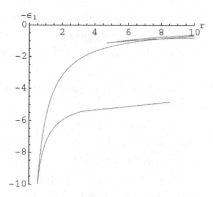

Fig. 3.13. Classical energy terms at $b = 3$ for $f = 2$.

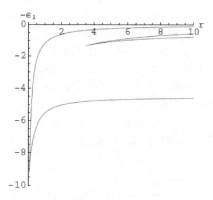

Fig. 3.14. Classical energy terms at $b = 3$ for $f = 0.1$.

The electric field causes several important new features compared to the zero-field case presented in Fig. 3.3. While at $f = 0$ there are three classical energy terms, the electric field brings up the fourth classical energy term. Indeed, let us take as an example the case of $f = 5$ at $b = 3$ presented in Fig. 3.12. There are four energy terms that we label as follows:

#1 — the lowest term;
#2 — the next term up (which has a V-type crossing with term 1);
#3 — the next term up;
#4 — the highest term (which has a V-type crossing with term 3).

We will use this labeling also while discussing all other plots: terms 1 and 2 will be those having the V-type crossing at the lower energy, terms 3 and 4 will be those having the V-type crossing at the higher energy,

At $f = 0$ term 2 is absent, but it appears at any non-zero value of $f$ — no matter how small. Actually, as $f$ approaches zero, this term behaves like $-f/r$, which is why it disappears at $f = 0$.

The existence of this additional term can be explained physically as follows. When $f = 0$, equilibrium of the orbital plane to the right of $Z'$ (i.e., for $w > 1$) does not exist, so that the values of $w_1$ and $w_3$ reduce to the ones presented in Sec. 3.3 and the right branch of $p(w)$ asymptotically goes to infinity when $w$ goes down to $w_3$. When an infinitesimal field $f$ appears, the right branch flips over positive infinity and ends up on the abscissa at $w_2 \to \infty$, thus enabling the whole region $w > 1$ for equilibrium. As the field grows, $w_2$ decreases. Physically, the force from the electric field at $w > 1$ balances out the Coulomb attraction of the $Z - Z'$ system on the left — the situation not possible for $f = 0$. This term is obtained by varying the parameter $w$ from 1 to $w_2$.

We emphasize that the above examples presented for $Z'/Z = 3$ represent a typical situation. In fact, *for any pair of $Z$ and $Z' \neq Z$*, at the presence of the electric field, there are four classical energy terms of the same symmetry for CRS.

Another important new feature caused by the electric field is X-type crossings of the classical energy terms. This kind of crossings and their physical consequences are discussed in the next section.

Figure 3.15 shows a magnified version of the energy terms 2, 3, and 4 at $b = 3$ for $f = 2$. Figure 3.16 shows a further magnified version of the energy terms 2 and 4 at $b = 3$ for $f = 2$. Compared to Fig. 3.13 for the same $b$ and $f$, in Figs. 3.15 and 3.16 we decreased the exhibited energy range, but increased the exhibited range of the internuclear distances $r$.

It is seen that term 2 has the X-type crossing with term 3 at $r = 7.8$ and the X-type crossing with term 4 at $r = 32$. The situation where there are two X-type crossings exists in a limited range of the

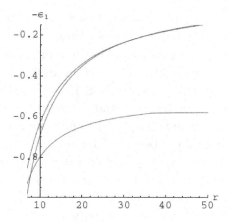

Fig. 3.15. Magnified plot of the classical energy terms 2, 3, and 4 at $b = 3$ for $f = 2$.

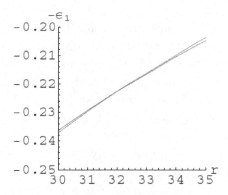

Fig. 3.16. Further magnified plot of the classical energy terms 2 and 4 at $b = 3$ for $f = 2$.

electric fields. For example, for $b = 3$:

— two X-type crossings exist at $1.31 < f < 2.4$;
— there are no X-type crossings at $f < 1.31$;
— there is one X-type crossing at $f > 2.4$ (the crossing of terms 2 and 3).

In Sec. 3.3 we explained that when $Z$ and $Z'$ differ significantly from each other, the V-type crossings occur between two classical energy terms that can be asymptotically labeled as $Z$- and $Z'$-terms, and that this situation *classically depicts charge exchange*. So, the first distinction caused by the electric field is an additional, second V-type crossing leading to charge exchange — compared to the zero-field case where there was only one such crossing. However, the second V-type crossing (the crossing of terms 1 and 2) occurs at the internuclear distance $r_{V2} \ll r_{V1}$, where $r_{V1}$ is the internuclear distance of the first V-type crossing (the crossing of terms 3 and 4). Therefore, the cross-section of the charge exchange corresponding to the second V-type crossing is much smaller than the corresponding cross-section for the first V-type crossing.

Now let us discuss the X-type crossing from the same point of view. When $Z$ and $Z'$ differ significantly from each other, the X-type crossing of terms 2 and 4 is the crossing of terms that can be asymptotically labeled as $Z$- and $Z'$-terms. Thus, this situation again *classically depicts charge exchange*. The most important is that this crossing occurs at the internuclear distance $r_{X1} \gg r_{V1} \gg r_{V2}$. Therefore, the cross-section of charge exchange due to this X-type crossing is much greater than the corresponding cross-sections of for the V-type crossings. This is the most fundamental physical consequence caused by the electric field: a *significant enhancement of charge exchange*.

When $Z$ and $Z'$ differ significantly from each other, the X-type crossing of terms 2 and 3 is the crossing of terms having the same asymptotic labeling: either both of them are $Z$-terms or both of them are $Z'$-terms. Therefore this second X-type crossing (at $r = r_{X2}$) does not correspond to charge exchange — rather it represents an *additional ionization channel*. Indeed, say, initially at $r \to \infty$, the electron resided on term 3 of the HI of the nuclear charge $Z$. As the distance between the charges $Z$ and $Z'$ decreases to $r = r_{X2}$, the electron can switch to term 2, which asymptotically corresponds to a near-zero-energy state (of the same HI of the nuclear charge $Z$) where the electron would be almost free. So, as the charges $Z$ and $Z'$ go away from each other, the system undergoes the ionization. Thus, another physical consequence caused by the electric field is the appearance of the additional ionization channel. This should have been expected since the electric field promotes the ionization of atomic and molecular systems.

## 3.6. Effects of the Screening by Plasma Electrons on Diatomic Rydberg Quasimolecules

Plasma screening of a test charge is a well-known phenomenon. For a hydrogen atom or a HI (an H-atom, for short), it is effected by replacing the pure Coulomb potential by a screened Coulomb potential which contains a physical parameter — the screening length $a$. For example, the Debye–Hückel (or Debye) interaction of an electron with the electronic shielded field of an ion of charge $Z$ is $U(R_0) = -(Ze^2/R_0)\exp(-R_0/a)$, where the so-called Debye radius is $a = (T/(4\pi e^2 N_e))^{1/2} \approx 6.90(T/N_e)^{1/2}$, where $N_e$ (cm$^{-3}$) and $T$ (K) are the electron density and temperature, respectively.

We study a two-Coulomb center (TCC) system with the charge $Z$ placed at the origin, and the $Oz$ axis is directed at the charge $Z'$, which is at $z = R$, the system being submerged in a plasma of a screening length $a$. We consider the circular orbits of the electron which are perpendicular to the internuclear axis and centered on the axis.

Two quantities, the energy $E$ and the projection $M$ of the angular momentum on the internuclear axis are conserved in this configuration. We use cylindrical coordinates to write the equations for both:

$$E = \frac{1}{2}\left(\dot{\rho}^2 + \rho^2\dot{\varphi}^2 + \dot{z}^2\right) - \frac{Z}{r}e^{-\frac{r}{a}} - \frac{Z'}{r'}e^{-\frac{r'}{a}}, \qquad (3.75)$$

$$M = \rho^2\dot{\varphi}, \qquad (3.76)$$

where $r$ and $r'$ are distances from the electron to $Z$ and $Z'$. The circular motion implies that $d\rho/dt = 0$; as the motion occurs in the plane perpendicular to the $z$-axis, $dz/dt = 0$. Further, expressing $r$ and $r'$ through $\rho$ and $z$, and taking $d\varphi/dt$ from Eq. (3.2), we have:

$$E = \frac{M^2}{2\rho^2} - \frac{Z}{\sqrt{\rho^2 + z^2}}e^{-\sqrt{\rho^2+z^2}/a} - \frac{Z'}{\sqrt{\rho^2 + (R-z)^2}}e^{-\sqrt{\rho^2+(R-z)^2}/a}.$$

$$(3.77)$$

With the scaled quantities

$$w = \frac{z}{R}, \quad p = \left(\frac{\rho}{R}\right)^2, \quad b = \frac{Z'}{Z}, \quad \varepsilon = -\frac{ER}{Z},$$

$$\ell = \frac{M}{\sqrt{ZR}}, \quad \lambda = \frac{R}{a}, \quad r = \frac{ZR}{M^2} \tag{3.78}$$

our energy equation takes the form below:

$$\varepsilon = \frac{e^{-\lambda\sqrt{w^2+p}}}{\sqrt{w^2+p}} + \frac{be^{-\lambda\sqrt{(1-w)^2+p}}}{\sqrt{(1-w)^2+p}} - \frac{\ell^2}{2p}. \tag{3.79}$$

We seek the equilibrium points by finding partial derivatives of $\varepsilon$ by the scaled coordinates $w$, $p$ and setting them equal to zero. This yields the following two equations:

$$\frac{we^{-\lambda\sqrt{w^2+p}}}{w^2+p}\left(\frac{1}{\sqrt{w^2+p}} + \lambda\right)$$

$$= \frac{b(1-w)e^{-\lambda\sqrt{(1-w)^2+p}}}{(1-w)^2+p}\left(\frac{1}{\sqrt{(1-w)^2+p}} + \lambda\right), \tag{3.80}$$

$$\frac{l^2}{p^2} = \frac{e^{-\lambda\sqrt{w^2+p}}}{w^2+p}\left(\frac{1}{\sqrt{w^2+p}} + \lambda\right)$$

$$+ \frac{be^{-\lambda\sqrt{(1-w)^2+p}}}{(1-w)^2+p}\left(\frac{1}{\sqrt{(1-w)^2+p}} + \lambda\right). \tag{3.81}$$

From the definitions of the scaled quantities (3.77), $\ell^2 = 1/r$ and $E = -(Z/R)\varepsilon$. Since $r = ZR/M^2$, then $E = -(Z/M)^2\varepsilon/r$, where $r = 1/\ell^2$ can be obtained by solving Eq. (3.81) for $\ell$. Thus, the scaled energy without explicit dependence on $R$ is $\varepsilon/r$, which we denote by $\varepsilon_1$. Using this, Eqs. (3.79), (3.80) and (3.81) can be transformed into the following three master equations for this configuration.

$$\varepsilon_1 = \left( \frac{p(1 + \lambda\sqrt{w^2 + p})e^{-\lambda\sqrt{w^2 + p}}}{(1 - w)(w^2 + p)^{3/2}} \right)^2$$

$$\times \left( \frac{(1 - w)(w^2 + p)}{1 + \lambda\sqrt{w^2 + p}} + \frac{w((1 - w)^2 + p)}{1 + \lambda\sqrt{(1 - w)^2 + p}} - \frac{p}{2} \right),$$

$$\tag{3.82}$$

$$r = \frac{(1 - w)(w^2 + p)^{3/2}e^{\lambda\sqrt{w^2 + p}}}{p^2(1 + \lambda\sqrt{w^2 + p})} \tag{3.83}$$

$$\frac{w(1 + \lambda\sqrt{w^2 + p})e^{-\lambda\sqrt{w^2 + p}}}{(w^2 + p)^{3/2}}$$

$$= \frac{b(1 - w)(1 + \lambda\sqrt{(1 - w)^2 + p})e^{-\lambda\sqrt{(1-w)^2 + p}}}{((1 - w)^2 + p)^{3/2}}. \tag{3.84}$$

The quantities $\varepsilon_1$ and $r$ now depend only on the coordinates $w$ and $p$ (besides the constant $\lambda$). Therefore, if we solve Eq. (3.84) for $p$ and substitute it into Eqs. (3.1) and (3.83), we obtain the parametric solution for the energy terms $\varepsilon_1(r)$ with the parameter $w$ for the given $b$ and $\lambda$.

Equation (3.84) does not allow an exact analytical solution for $p$. Therefore, we will use an approximate analytical method.

Figure 3.17 shows the contour plot of this equation for $b = 3$ and $\lambda = 0.1$.

As in Sec. 3.5, the plot has two branches, the left one spanning from $w = 0$ to $w = w_1$, and the right one from the asymptote $w = w_3$ to $w = 1$. Here $w_1$ is a solution of the equation

$$(1 - w_1)^2(1 + \lambda w_1)e^{\lambda(1 - 2w_1)} = bw_1^2(1 + \lambda(1 - 2w_1)) \tag{3.85}$$

in the interval $0 < w_1 < 1$, and $w_3$ does not depend on $\lambda$ and equals $b/(1 + b)$ — the same as for $\lambda = 0$, i.e., for the no-screening case described in Sec. 3.3. As $\lambda$ increases, $w_1$ and the $p$-coordinate of the maximum of the left branch increase, but the general shape of both

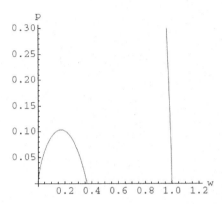

Fig. 3.17. Contour plot of equation (3.84) for $b = 3$ and $\lambda = 0.1$.

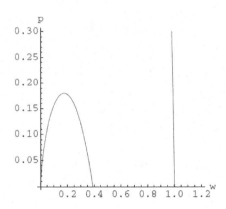

Fig. 3.18. Contour plot of Eq. (3.84) for $b = 3$ and $\lambda = 2$.

curves is preserved. Figure 3.18 shows the plot for a relatively strong screening: $\lambda = 2$.

An approximation was made for small values of $\lambda$. Approximating Eq. (3.84) in the first power of $\lambda$, we obtain the expression involving only the second and higher powers of $\lambda$. Therefore, an attempt was made using the value of $p(w)$ for $\lambda = 0$ presented in Sec. 3.3, which we shall denote as $p_0$; it is equal to the square of the right side of Eq. (3.11). Further, taking the higher powers of $\lambda$ into account, we obtained the next-order approximation for $p(w)$:

$$p(w) = p_0 + \frac{\lambda^2}{6}(1 - 2w)\left(1 + (1 - 2w)\left(\frac{w^{2/3} + b^{2/3}(1 - w)^{2/3}}{w^{2/3} - b^{2/3}(1 - w)^{2/3}}\right)\right),$$
(3.86)

where

$$p_0 = \frac{w^{2/3}(1 - w)^{4/3} - b^{2/3}w^2}{b^{2/3} - w^{2/3}(1 - w)^{-2/3}}$$
(3.87)

is the zero-$\lambda$ value as in the square of the right side of Eq. (3.11). Equation (3.85) can be approximated by substituting $1 + \lambda(1 - 2w_1)$ in place of $\exp(\lambda(1 - 2w_1))$, which will render it a fourth-degree polynomial in $w_1$.

Substituting Eq. (3.86) into Eqs. (3.82) and (3.83), we obtain the approximate parametric solution for the energy terms $-\varepsilon_1(r)$ by running the parameter $w$ on the intervals $0 < w < w_1$ and $w_3 < w < 1$. Empirically, by comparison with the numerical results, it was found that using the value of $p$ from Eq. (3.87) on the $0 < w < w_1$ range and from Eq. (3.86) on the $w_3 < w < 1$ range yields the best approximate results. Figures 3.19 and 3.20 show the approximate terms for $b = 3$ and the screening parameter $\lambda = 0.1$ and $\lambda = 0.2$, respectively,

A numerical solution has also been made. It confirmed that the analytical solution was a good approximation for $\lambda < 0.3$. Figures 3.21 and 3.22 show the numerically calculated terms for the screening parameter $\lambda = 0.2$ and $\lambda = 2$, respectively.

Several properties of these energy terms have been studied. We note that in case of small or moderate $\lambda$, we observe four terms, both pairs of which have a V-type crossing. As an example, let us take the plot of the terms for $\lambda = 0.2$ in Fig. 3.21 and number the lowest term 1 and the highest term 2; the remaining terms will be numbered 3 and 4, from the lower one to the higher one. Therefore, terms 1 and 2 and terms 3 and 4 undergo V-type crossings, to which we shall refer as V12 and V34. Using a small-$\lambda$ approximation by choosing Eq. (3.87) as the $p(w)$ solution for the parametric energy

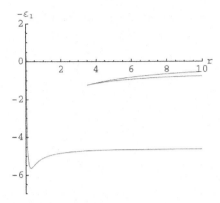

Fig. 3.19. Approximate classical energy terms for $b = 3$ at the screening parameter $\lambda = 0.1$.

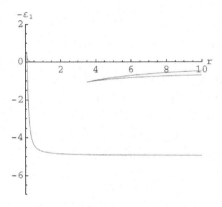

Fig. 3.20. Approximate classical energy terms for $b = 3$ and the screening parameter $\lambda = 0.2$.

terms (essentially, a zero-$\lambda$ approximation), we substitute Eq. (3.87) into Eq. (3.83) and obtain the following result.

$$r = \frac{(1 - 2w)^{3/2}\sqrt{b^{2/3} - \left(\frac{w}{1-w}\right)^{2/3}}}{w^3 \left(b^{2/3} - \left(\frac{1-w}{w}\right)^{4/3}\right)^2}. \tag{3.88}$$

For a given $b$, the terms 3 and 4 are produced by varying $w$ between 0 and $w_1$. The V34 crossing occurs at the value of $w$ where

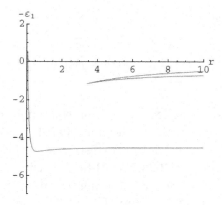

Fig. 3.21. Numerically calculated classical energy terms for $b = 3$ and $\lambda = 0.2$.

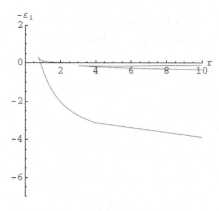

Fig. 3.22. Numerically calculated classical energy terms for $b = 3$ and $\lambda = 2$.

$r(w)$ reaches its minimum value $r_{\min}$ — since these terms do not exist at $r < r_{\min}$. Therefore, setting the derivative $dr/dw$ to zero, we obtain the equation whose solution for $w$ in the range $0 < w < w_1$ gives us the point on the parametric axis which produces the V34 crossing.

$$9w^{\frac{4}{3}}(1 - w)^{\frac{4}{3}}\left(w^{\frac{4}{3}} + b^{\frac{4}{3}}(1 - w)^{\frac{4}{3}}\right)$$
$$= b^{\frac{2}{3}}(1 - 4w + 22w^2 - 36w^3 + 18w^4). \qquad (3.89)$$

This equation has no dependence on $\lambda$ and is therefore equivalent to the Coulomb-potential case described in Sec. 3.3. Therefore, the following analytical solution exists for Eq. (3.89):

$$w_{V34} = \frac{1}{1 + \left(b^{1/3} + \frac{(b-1)^{1/3}}{b^{1/6}}((\sqrt{b}+1)^{1/3} + (\sqrt{b}-1)^{1/3})\right)^{3/2}}.$$

(3.90)

Substituting it into Eq. (3.83) and using the numerical solution for $p$ of Eq. (3.84), we obtain the semi-analytical dependence $r_{V34}(\lambda)$ for a given $b$. Figure 3.23 shows the corresponding plot for $b = 3$.

Since the plot in Fig. 3.23 was obtained using a zero-$\lambda$ approximation for the point of the V34 crossing, we also graphed this dependence numerically point by point. Figure 3.24 below for the same $b$ shows that in relation to terms 3 and 4, this approximation works well even for relatively large values of $\lambda$.

The energy of the V34 crossing can be obtained semi-analytically by substituting the numerical solution for $p$ of Eq. (3.84) into

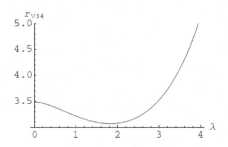

Fig. 3.23. Semi-analytical plot of the scaled internuclear distance $r_{V34}(\lambda)$ at the V-type crossing of terms 3 and 4 for $b = 3$.

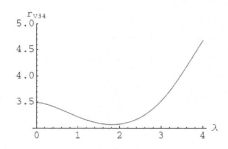

Fig. 3.24. Numerically calculated plot of the same dependence as in in Fig. 3.23.

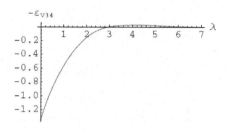

Fig. 3.25. Numerically calculated plot of the scaled energy $-\varepsilon_{V34}(\lambda)$ at the V-type crossing of terms 3 and 4 for $b = 3$.

Eq. (3.82), and by further substituting $w_{V34}$ from Eq. (3.90) into the resulting formula. As $\lambda$ grows, the energy of the crossing also grows and at a relatively large $\lambda$ becomes positive. A numerically calculated graph can also be made in a fashion similar to Fig. 3.24. Figure 3.25 presents it for $b = 3$.

It is seen that the energy of the V34 crossing becomes positive after $\lambda = 2.96$, then has a maximum, and later asymptotically approaches zero. At another value of $b = 4/3$, the V34 crossing reaches zero energy at $\lambda = 2.13$.

The screening also affects the shape of the terms 3 and 4. Term 3, whose energy increases as $r$ increases, becomes nearly horizontal at energy $-0.5$ at a certain value of $\lambda = \lambda_{cr}$. At $\lambda > \lambda_{cr}$, its energy decreases with $r$. For $b = 3$, $\lambda_{cr}$ is about 1.1; for $b = 4/3$, it is about 0.7. The corresponding plot for $b = 3$ is shown in Fig. 3.26.

For V12 crossing, the small-$\lambda$ approximation is not applicable since this crossing is not observed at $\lambda = 0$. Therefore, only numerical methods are used. A situation of particular interest is the behavior of term 1 at very small $r$, because as $r \to 0$ it corresponds to the energy of the hydrogenic ion of the nuclear charge $Z + Z'$. The plasma screening in this limit was studied in paper [3.25]. The point with the smallest $r$ is the V12 crossing. We compare the dependence of the electronic energy on the screening parameter $\lambda$ presented in [3.25] with the limiting case $r \to 0$ in our situation. Since in [3.25], the calculation was performed for a single Coulomb center $Z$, we have to re-scale the quantities to make an adequate comparison. The electronic energies are related as $\varepsilon_1^{(TCC)} = (1 + b)^2 \varepsilon_1^{(OCC)}$, where

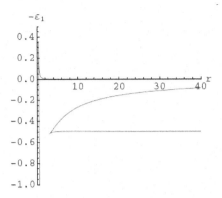

Fig. 3.26. Classical energy terms for $b = 3$ at $\lambda = 1.1$; term 3 is nearly constant at the scaled energy $-\varepsilon_1 = -0.5$.

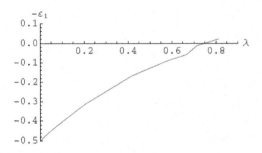

Fig. 3.27. Dependence of the scaled energy at $r \to 0$ on the screening factor $\lambda$ for $b = 3$ in the TCC case.

OCC stands for "one Coulomb center". While the scaling for the screening parameter in the OCC case naturally did not include the internuclear distance $R$, the relation between the screening parameters $\lambda^{(TCC)}$ and $\lambda^{(OCC)}$ does include the scaled internuclear distance $r$: $\lambda^{(TCC)} = r(1 + b)\lambda^{(OCC)}$. Taking this into account, we plot in Fig. 3.27 for the TCC case, the dependence of the scaled energy at $r \to 0$ on the screening factor $\lambda$ (we note that the inflection point at $r = 0.67$ is an artifact).

Figure 3.28 presents is the same dependence obtained in paper [3.25] for the OCC case.

Another aspect of this problem worth studying is the internuclear potential. Previously its properties were studied in paper [3.20] for

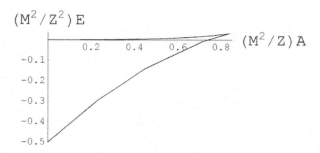

Fig. 3.28. Plot of the energy of the electron in an OCC system versus the scaled screening factor.

the same system with $\lambda = 0$ and a magnetic field parallel to the internuclear axis — as presented in this book in Sec. 3.4. Particularly, the magnetic field created a deep minimum in the internuclear potential, which stabilized the nuclear motion and transformed a Rydberg quasi-molecule into a real molecule. Here we investigate the effect of the screening on the internuclear potential. Its form in atomic units is

$$U_{\text{int}} = \frac{ZZ'}{R} + E, \tag{3.91}$$

where $E$ is the electronic energy. Using the scaled quantities from Eq. (3.78), we get the scaled internuclear potential in the form

$$u = \frac{bZ}{r} - \varepsilon_1, \tag{3.92}$$

where $U_{\text{int}} = (Z/L)^2 u$. By plotting its dependence on $r$, we found out that in cases of $Z > 1$ the screening tends to flatten the minimum, producing the effect opposite to the one of the magnetic field. This can be seen by comparing the plots of $u(r)$ in the case of $Z = 2$, $b = 2$ for $\lambda = 0$ (Fig. 3.29) and $\lambda = 0.3$ (Fig. 3.30).

The screening increases the potential of the point of intersection of the two branches; the upper branch, which has a very shallow minimum at $\lambda = 0$, loses it as $\lambda$ increases.

A completely different behavior is observed for $Z = 1$. A small $\lambda$ creates a deep minimum in the upper branch of the potential. This can be seen by comparing the plots of $u(r)$ in the case of $Z = 1$, $b = 2$ for $\lambda = 0$ (Fig. 3.31) and $\lambda = 0.3$ (Fig. 3.32).

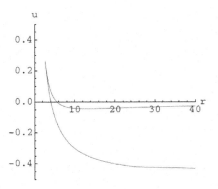

Fig. 3.29. Scaled internuclear potential versus the scaled internuclear distance for $Z = 2$, $Z' = 4$, $\lambda = 0$.

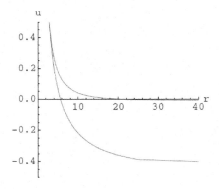

Fig. 3.30. Same as in Fig. 3.29, but for the screening factor $\lambda = 0.3$.

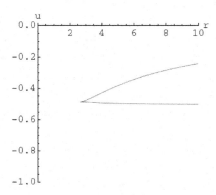

Fig. 3.31. Scaled internuclear potential versus the scaled internuclear distance for $Z = 1$, $Z' = 2$, $\lambda = 0$.

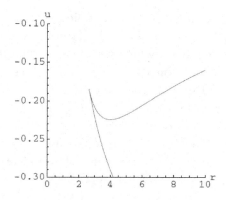

Fig. 3.32. Same as in Fig. 3.31, but for the screening factor $\lambda = 0.3$.

Figures 3.29–3.32 illustrate the principal difference between the cases of $Z = 1$ and $Z > 1$, which is a counterintuitive result. Namely, the plasma screening stabilizes the nuclear motion for the case of $Z = 1$, but destabilize it for $Z > 1$.

## 3.7. Applications to CL in Plasmas

CL is a fundamental concept of atomic physics in plasmas. It refers to the fact that the energy required to ionize is reduced compared to its value in vacuum. This is because highly-excited states above certain threshold disappear from the discrete spectrum. The higher the plasma density, the more significant CL becomes. It defines the existing energy states and affects their absorption and emission properties. CL is important for inertial fusion, X-ray lasers, astrophysics, and other applications of high-density plasmas. It is usually included in most comprehensive simulations of atomic physics in plasmas. CL plays a key role in calculations of the equation of state, partition function, bound-free opacities, and other collisional atomic transitions in plasmas.

CL was studied for over 50 years — see, e.g., books/reviews [3.48–3.52] and references therein. Calculations of CL evolved from ion sphere models to dicenter models of the plasma state [3.48, 3.53–3.58]. One of such theories — a percolation theory [3.48, 3.55] — calculated CL defined as an absolute value of energy at which an electron becomes bound to a macroscopic portion of plasma ions (a quasi-ionization). In 2001 we derived analytically the value of CL in the true-ionization channel, which was disregarded in the percolation theory: a quasimolecule, consisting of the two ion centers plus an electron, can get ionized in a true sense of this word before the electron would be shared by more than two ions [3.21]. In other words, the distinction between the quasi-ionization (i.e., the percolation theory) and the true ionization is the following. The basic process in the percolation theory is the tunneling of the electron between the two adjacent ionic potential wells. But this means that the electron is still a part of the quasimolecule consisting of the two adjacent ions and the electron, i.e., the electron is still bound. However, it is well-known from molecular physics that any molecule or quasi-molecule can actually lose an electron, i.e., get truly ionized. This represents the true-ionization channel of CL — in addition to the quasi-ionization channel of the percolation theory.

It was also shown in paper [3.21] that, whether the electron is bound primarily by the smaller or by the larger out of two positive charges $Z$ and $Z'$, makes a dramatic qualitative and quantitative difference for this ionization channel.

Here we study how three different factors affect the value of CL in the ionization channel. The first factor is the screening by plasma electrons that was disregarded in paper [3.21]. The second factor is an electric field. It represents (in a model way) a quasistatic ion microfield due to contributions of all ions except the two ions included in the dicenter. The third factor is a magnetic field. Relevant applications of the latter include (but are not limited to) laser fusion, where a strong magnetic field can be generated in the process of the laser–plasma interaction [3.59, 3.60], and powerful Z-pinches used for producing X-ray and neutron radiation, ultra-high pulsed magnetic fields, and for X-ray lasing (see, e.g., [3.61]). We show that the screening and the magnetic field decrease the value of CL, inhibiting the ionization, while the electric field increases the value of CL, promoting the ionization.

Our analysis of the stability of the electronic motion shows results similar to those presented in Sec. 3.2. Namely, term 3 corresponds to a stable motion, while term 4 — to an unstable motion. So, *the crossing point of terms 3 and 4 corresponds to the transition from the stable motion to the unstable motion, leading the electron to the zero energy (i.e., to the free motion) along term 4, which constitutes the ionization of the molecule.*

Therefore, we arrive at the following. For the ionization of the HI of the nuclear charge $Z_{\min}$ perturbed by the charge $Z_{\max}$, it is sufficient to reach the scaled energy $\varepsilon_c(b) = \varepsilon[w_{V34}(b), b] < 0$. At that point, the electron switches to the unstable motion and the radius of its orbit increases without a limit. This constitutes CL by the amount of $Z\langle 1/R\rangle|\varepsilon\,(w_{V34}(b), b)|$, where $\langle 1/R\rangle$ is the value of the inverse distance of the nearest neighbor ion from the radiating ion averaged over the ensemble of perturbing ions. Thus, obtaining CL in the ionization channel requires calculations of the scaled energy $\varepsilon$ at the crossing point $w_{V34}$ of terms 3 and 4.

Now we study the effect of plasma screening on CL in the ionization channel. The no-screening case of CL, i.e., for the screening factor $\lambda = 0$, was presented in paper [3.21]. (We remind readers that $\lambda = R/a$ was defined in Eq. (3.78), where $a$ is the Debye radius: $a = (T/(4\pi e^2 N_e))^{1/2} \approx 6.90[T(K)/N_e(\mathrm{cm}^{-3})]^{1/2})$. Particularly, the dependence of the scaled value of CL $\varepsilon_c(b) = \varepsilon(w_{V34}(b), b) = \Delta E/(Z\langle 1/R\rangle)$ versus $b$ was plotted in [3.21] in the double-logarithmic scale, where $\varepsilon = -ER/Z$ (see Eq. (3.65)) and $w_{V34}$ is given by Eq. (3.90). This plot is presented in Fig. 3.33; "log $x$" stands for "$\log_{10}x$".

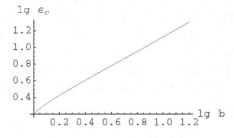

Fig. 3.33. The CL value $\varepsilon_c$ versus $b$ in the double logarithmic scale for the no-screening case ($\lambda = 0$).

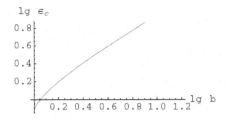

Fig. 3.34. Same as in Fig. 3.33, but for the screening factor $\lambda = 0.5$.

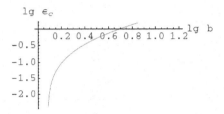

Fig. 3.35. Same as in Fig. 3.33, but for the screening factor $\lambda = 2$.

Figures 3.34 and 3.35 show $\varepsilon_c(b)$ for the screening factors $\lambda = 0.5$ and $\lambda = 2$. We used numerical values for $w_{V34}$ for increasing the precision.

From Figs. 3.34 and 3.35 it is seen that the plasma screening decreases the value of CL in the ionization channel. Also, starting from about $\lambda = 1.7$, we observe the "cutoff" value of $b > 1$, below which $\varepsilon_c$ becomes negative, i.e., the electron energy at $w_{V34}$ becomes positive. This means that there is no more CL in this ionization channel — instead, the continuum becomes *higher* than for the isolated HI of the nuclear charge $Z$. This effect cannot be observed in the logarithmic graphs above because the cutoff value of energy (equal to zero) corresponds to $\log \varepsilon_c = -\infty$. Figure 3.36 presents the standard, non-logarithmic plot of $\varepsilon_c(b)$ for $\lambda = 3$, the plot clearly showing this effect. Namely, it is seen, e.g., that there is no continuum "lowering" for $b = 2$ and $b = 3$.

In the next two sections we study the effects of electric and magnetic fields on CL in the ionization channel. We will show that the magnetic field decreases the value of CL, similar to the effect of the

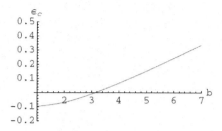

Fig. 3.36. The CL value versus $b$ for the screening factor $\lambda = 3$.

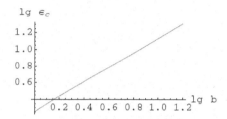

Fig. 3.37. The CL value $\varepsilon_c$ versus $b$ in the double logarithmic scale for the scaled electric field $f = 0.1$.

plasma screening, while the electric field increases the value of CL. Physically this is because the magnetic field inhibits the ionization, while the electric field promotes the ionization.

For studying the effect of the electric field on CL in the ionization channel, we use the value of the scaled energy of the electron in the TCC system given in Eq. (3.66) with the substitution of the numeric or approximate analytical solution for $p$ from Eq. (3.71) into Eqs. (3.66) and (3.68). Then we substitute $\ell$ from Eq. (3.68) into Eq. (3.66) and obtain the dependence of the scaled energy on the scaled coordinate $w$ for the case where the electric field was parallel to the internuclear axis. Thereafter we numerically find the point on the $w$-axis corresponding to the V34 crossing for a given value of the scaled electric field $f$ and substitute it into the formula for the scaled energy, thus obtaining the critical energy, which is the value of CL.

Figures 3.37–3.39 present plots of log $\varepsilon_c$ versus log $b$ for three values of the scaled electric field $f = FR^2/Z$: $f = 0.1$ (weak field), $f = 1$ (moderate field), and $f = 10$ (strong field), respectively.

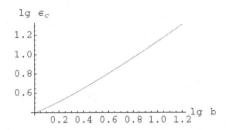

Fig. 3.38. Same as in Fig. 3.37, but for $f = 1$.

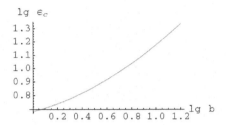

Fig. 3.39. Same as in Fig. 3.37, but for $f = 10$.

It is seen that CL increases as the electric field increases. This is expected because the electric field promotes ionization.

Finally we study the effect of the magnetic field on CL in the ionization channel for the case where the magnetic field **B** is parallel to the internuclear axis. Using the $\gamma$-parametrization instead of the $w$-parametrization ($\gamma$ being defined by Eq. (3.41)), the expressions for the scaled energy $\varepsilon$ (from Eqs. (3.48) and (3.58)) and for the scaled internuclear distance $r$ (from Eq. (3.57)) can be represented as follows:

$$\varepsilon = \frac{(\gamma^4 - 2\gamma + b^{2/3}(2\gamma^3 - 1))\sqrt{(\gamma^3 + 1)(b^{2/3}\gamma^2 - 1)}}{2\gamma(\gamma^3 - 1)^{3/2}}$$
$$+ \frac{\gamma^2(b^{2/3} - \gamma^4)}{(\gamma^3 + 1)^2(b^{2/3}\gamma^2 - 1)}$$

$$\times \omega \left( \omega + \sqrt{\omega^2 + \frac{(\gamma^3 + 1)^{5/2}(b^{2/3}\gamma^2 - 1)^{3/2}}{\gamma^3(\gamma^3 - 1)^{3/2}}} \right), \qquad (3.93)$$

$$r = \frac{(\gamma^3 + 1)^4 (b^{2/3}\gamma^2 - 1)^2}{\gamma^4(b^{2/3} - \gamma^4)^2 \left( \omega^2 + \frac{(\gamma^3+1)^{5/2}(b^{2/3}\gamma^2-1)^{3/2}}{\gamma^3(\gamma^3-1)^{3/2}} \right)}. \qquad (3.94)$$

We remind readers that $\omega = \Omega M^3/Z^2 = BM^3/(2cZ^2)$ is the scaled magnetic field, $\Omega$ being the Larmor frequency (see Eqs. (3.46) and (3.47)).

To find the point of the V34 crossing, we take the derivative of $r$ by $\gamma$ and set it equal to zero. The numerical solution for this equation determines the value of $\gamma$ corresponding to the minimum of $r$ (for given $b$ and $\omega$) which corresponds to the crossing, as already discussed in the previous sections of this chapter. Substituting this value of $\gamma$ into Eq. (3.93) for the scaled energy, we obtain $\varepsilon_c(b,\omega)$, i.e., the dependence of the scaled value of CL on $b$ for a given scaled magnetic field $\omega$.

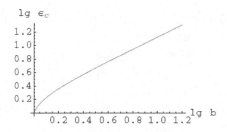

Fig. 3.40. The CL value $\varepsilon_c$ versus $b$ in the double logarithmic scale for the scaled magnetic field $\omega = 0.5$.

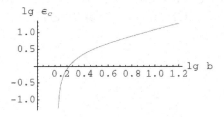

Fig. 3.41. Same as in Fig. 3.40, but for the scaled magnetic field $\omega = 2$.

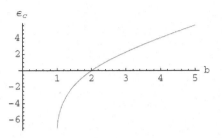

Fig. 3.42. The CL value $\varepsilon_c$ versus $b$ for the scaled magnetic field $\omega = 2.8$.

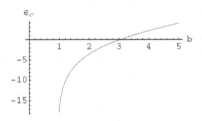

Fig. 3.43. Same as in Fig. 3.42, but for the scaled magnetic field $\omega = 4.3$.

Figures 3.40–3.43 present several plots of the value of the CL $\varepsilon_c$ versus $b$ for selected values of $\omega$.

It is seen that the effect of the magnetic field on CL in the ionization channel is similar to the effect of the plasma screening. Namely, the magnetic field decreases the value of CL.

Similarly to the case of the plasma screening, as the scaled magnetic field $\omega$ increases, there appear "cutoff" values of $b$, below which there is no CL, but rather continuum *raising*. For example, for $\omega = 2.8$, the value of CL vanishes at $b = 2$: the values of $b$ corresponding to CL start at $b > 2$, as shown in Fig. 3.42. Another example: for $\omega = 4.3$, CL starts at $b > 3$, as shown in Fig. 3.43.

Chapter 4

# Classical Stationary States and non-Einsteinian Time Dilation: Generalized Hamiltonian Dynamics (GHD)

## 4.1. GHD for the Motion in the Coulomb Potential

In the middle of the 20th century, Dirac developed a generalized Hamiltonian dynamics (GHD) [4.1–4.3]. The conventional Hamiltonian dynamics is based on the assumption that momenta are independent functions of velocities. Dirac [4.1–4.3] analyzed a more general situation where *momenta are **not** independent functions of velocities*. Physically, the GHD is a purely classical formalism for systems having constraints; it incorporates the constraints into the Hamiltonian. Dirac designed the GHD specifically for applications to quantum field theory [4.4].

Here we apply the GHD to atomic and molecular physics by choosing integrals of the motion as the constraints. After a general description of our formalism, we consider hydrogenic atoms as an example. We show that this purely classical formalism allows deriving the expression for the energy of the classical non-radiating states coinciding with the corresponding quantal expression. In the latter derivation we employed two fundamental experimental facts, but did not "forcefully" quantize any physical quantity describing the atom. In particular, this amounts to classically deriving the Bohr's postulate on the quantization of the angular momentum (instead of postulating it).

We show that our classical formalism can be applied to all atomic and molecular systems. We also discuss advantages of our formalism over classical and semiclassical models employed in chemical physics for describing electronic degrees of freedom.

Dirac [4.1–4.3] considered a dynamical system of $N$ degrees of freedom characterized by generalized coordinates $q_n$ and velocities $v_n = dq_n/dt$, where $n = 1, 2, \ldots, N$. If the Lagrangian of the system is

$$L = L(q, v), \tag{4.1}$$

then momenta are defined as

$$p_n = \partial L/\partial v_n. \tag{4.2}$$

Each of the quantities $q_n$, $v_n$, $p_n$ can be varied by $\delta q_n$, $\delta v_n$, $\delta p_n$, respectively. The latter small quantities are of the order of $\varepsilon$, the variation being worked to the accuracy of $\varepsilon$. As a result of the variation, Eq. (4.2) would not be satisfied any more, since their right side would differ from the corresponding left side by a quantity of the order of $\varepsilon$.

Further, Dirac distinguished between two types of equations. To one type belong equations such as Eq. (4.2), which do not hold after the variation (he called them "weak" equations). Below for weak equations, following Dirac, we use an equality sign $\cong$ different from the usual. Another type constitute equations such as Eq. (4.1), which holds exactly even after the variation (he called them "strong" equations).

If quantities $\partial L/\partial v_n$ are *not* independent functions of velocities, one can exclude velocities $v_n$ from Eq. (4.2) and obtain one or several weak equations

$$\phi(q, p) \cong 0, \tag{4.3}$$

containing only $q$ and $p$. In his formalism, Dirac [4.1–4.3] used the following complete system of independent equations of the type (4.3):

$$\phi_m(q, p) \cong 0, \quad (m = 1, 2, \ldots, M). \tag{4.4}$$

Here the word "independent" means that neither of the $\phi$'s can be expressed as a linear combination of the other $\phi$'s with coefficient depending on $q$ and $p$. The word "complete" means that any function of $q$ and $p$, which would become zero allowing for Eq. (4.2) and which would change by $\varepsilon$ under the variation, should be a linear combination of the functions $\phi_m(q,p)$ from Eq. (4.4) with coefficients depending on $q$ and $p$.

Finally, proceeding from the Lagrangian to a Hamiltonian, Dirac [4.1–4.3] obtained the following central result:

$$H_{\mathrm{g}} = H(q,p) + u_m \phi_m(q,p) \qquad (4.5)$$

(here and below, the summation over a twice repeated suffix is understood). Equation (4.5) is a strong equation expressing a relation between the generalized Hamiltonian $H_{\mathrm{g}}$ and the conventional Hamiltonian $H(q,p)$. Quantities $u_m$ are coefficients to be determined. Generally, they are functions of $q$, $v$, and $p$; by using Eq. (4.2), they could be made functions of $q$ and $p$. It should be emphasized that $H_{\mathrm{g}} \cong H(q,p)$ would be only a weak equation — in distinction to Eq. (4.5).

Equation (4.5) shows that the Hamiltonian is not uniquely determined, because a linear combination of $\phi$'s may be added to it. Equation (4.4) are called *constraints*. The above distinction between constraints (i.e., weak equations) and strong equations can be reformulated as follows.

Constraints must be employed in accordance to certain rules. Constraints can be added. Constraints can be multiplied by factors (depending on $q$ and $p$), but only on the left side, so that these factors must not be used inside Poisson brackets.

If $f$ is some function of $q$ and $p$, then $df/dt$ (i.e., a general equation of motion) in the Dirac's GHD is

$$df/dt \cong [f, H] + u_m[f, \phi_m], \qquad (4.6)$$

where $[f, g]$ is the usual Poisson bracket. Substituting $\phi_{m'}$ in Eq. (4.6) instead of $f$ and taking into account Eq. (4.4), one obtains:

$$[\phi_{m'}, H] + u_m[\phi_{m'}, \phi_m] \cong 0. \quad (m' = 1, 2, \ldots, M). \qquad (4.7)$$

These consistency conditions allow determining the coefficients $u_m$.

We note that the GHD was designed by Dirac specifically for applications to quantum field theory [4.17], that is, for the purpose totally different from the purpose of our work.

It is well-known that many dynamical systems possess *integrals of the motion* other than the energy. However, in our view, their role in atomic and molecular physics was not fully appreciated yet. Below we suggest using integrals of the motion as the constraints in the GHD.

We consider a classical atomic or molecular system of $N$ degrees of freedom, possessing $M$ classical integrals of the motion $A_m(q,p)$, $m = 1, 2, \ldots, M$. We write the generalized Hamiltonian in the form (see Eqs. (4.4) and (4.5))

$$H_g = H(q,p) + u_m\{A_m(q,p) - A_{0m}\}, \quad A_{0m} = \text{const.} \quad (4.8)$$

Here $A_{0m}$ is the value of $A_m(q, \text{i})$ in a particular state of the motion, so that in this state

$$A_m(q,p) - A_{0m} \cong 0. \quad (4.9)$$

Since the quantities $A_m(q,p)$ are integrals of the motion, their Poisson bracket with $H(q,p)$ vanishes and the consistency condition (4.7) reduces to the form

$$u_m[A_{m'}, A_m] \cong 0. \quad (m' = 1, 2, \ldots, M). \quad (4.10)$$

Equation (4.10) allows determining the coefficients $u_m$.

For a hydrogenic atom of the nuclear charge $Z$, the integrals of the motion (other than the energy) are the angular momentum $\mathbf{L} = \mathbf{r} \times \mathbf{p}$ and the Runge–Lenz vector [4.5] $\mathbf{A}(\mathbf{r}, \mathbf{p}) = \{\mathbf{r}p^2 - \mathbf{p}(\mathbf{r} \cdot \mathbf{p})\}/(\mu Z e^2) - \mathbf{r}/r$, where $\mu$ is the reduced mass. Therefore, the generalized Hamiltonian can be presented in the form

$$H_g = p^2/(2\mu) - Ze^2/r + \mathbf{u} \cdot (\mathbf{r} \times \mathbf{p} - \mathbf{L}_0)$$
$$+ \mathbf{w} \cdot (\mathbf{A}(\mathbf{r}, \mathbf{p}) - \mathbf{A}_0), \quad (4.11)$$

where $\mathbf{L}_0$, $\mathbf{A}_0$, and the energy $H_0$ are connected by the well-known relation [4.5]

$$L_0^2 = \mu Z^2 e^4 (A_0^2 - 1)/(2H_0). \quad (4.12)$$

The consistency conditions $[\mathbf{r} \times \mathbf{p}, H_g] \cong 0$, $[\mathbf{A}(\mathbf{r}, \mathbf{p}), H_g] \cong 0$ result into the following equations for the unknown vector-coefficients $\mathbf{u}$ and $\mathbf{w}$:

$$\mathbf{u} \times \mathbf{L}_0 + \mathbf{w} \times \mathbf{A}_0 \cong 0, \quad \mathbf{u} \times \mathbf{A}_0 - 2\mathbf{w} \times \mathbf{A}_0 H_0/(\mu Z^2 e^4) \cong 0. \quad (4.13)$$

We seek $\mathbf{u}$ and $\mathbf{w}$ in terms of their projections on the following three mutually orthogonal directions in space $\mathbf{A}_0$, $\mathbf{M}_0$, $\mathbf{A}_0 \times \mathbf{M}_0$:

$$\mathbf{u} = a_1 \mathbf{A}_0 + a_2 \mathbf{M}_0 + a_3 \mathbf{A}_0 \times \mathbf{M}_0;$$
$$\mathbf{w} = b_1 \mathbf{A}_0 + b_2 \mathbf{M}_0 + b_3 \mathbf{A}_0 \times \mathbf{M}_0. \quad (4.14)$$

Substituting Eq. (4.14) in Eq. (4.13), we find:

$$a_3 = b_3 = 0, \quad a_1 = b_2, \quad a_2 = -2b_1 H_0/(\mu Z^2 e^4). \quad (4.15)$$

Now, using the generalized Hamiltonian $H_g$, in which only two coefficients $b_1$ and $b_2$ remain unknown, we calculate equations of the motion:

$$d\mathbf{r}/dt = [\mathbf{r}, H_g] = \{1 + b_1(A_0^2 - 1)/(2H_0)\}\mathbf{p}/\mu,$$
$$d\mathbf{p}/dt = [\mathbf{p}, H_g] = -\{1 + b_1(A_0^2 - 1)/(2H_0)\}Ze^2\mathbf{r}/r^3. \quad (4.16)$$

First of all, since the equations of the motion do not depend on $b_2$, we can set $b_2 = 0$ in $H_g$ without any loss of generality. Second, it should be emphasized that the coefficient $b_1$ in Eq. (16) can still be a function of the scalar invariants $H_0$, $A_0$. Therefore, to simplify notations we denote:

$$b_1(A_0^2 - 1)/(2H_0) \equiv B(H_0, A_0). \quad (4.17)$$

In terms of this unknown function $B(H_0, A_0)$, the generalized Hamiltonian and the equations of the motion take the form:

$$H_g = p^2/(2\mu) - Ze^2/r + 2B(H_0, A_0)H_0$$
$$\times \{\mathbf{M}_0 \cdot (\mathbf{r} \times \mathbf{p})/M_0^2 - (1 - \mathbf{A}_0 \cdot \mathbf{A}(\mathbf{r}, \mathbf{p}))/(1 - A_0^2)\}, \quad (4.18)$$

$$d\mathbf{r}/dt = \{1 + B(H_0, A_0)\}\mathbf{p}/\mu,$$
$$d\mathbf{p}/dt = -\{1 + B(H_0, A_0)\}Ze^2\mathbf{r}/r^3. \quad (4.19)$$

The equations of the motion (4.19) differ from their conventional form only by the factor $\{1+B(H_0, A_0)\}$. Therefore, if we would make a transformation of time

$$t' = \{1 + B(H_0, A_0)\}t, \tag{4.20}$$

then with respect to the new time $t'$, the equations of the motion would be formally brought back to their conventional form.

Thus we come to the following *central point*. In the above generalized formalism, the trajectory of the atomic electron remains the same as in the conventional formalism. However, the generalized period $T_{\mathrm{g}}$ and the generalized frequency $\omega_{\mathrm{g}}$ differ from their conventional values $T_0$ and $\omega_0$ as follows:

$$T_{\mathrm{g}} = T_0/|1 + B(H_0, A_0)|, \tag{4.21}$$
$$\omega_{\mathrm{g}} = \omega_0|1 + B(H_0, A_0)| = |1 + B(H_0, A_0)||2H_0|^{3/2}/D^{1/2},$$
$$D = \mu Z^2 e^4. \tag{4.22}$$

It is seen from Eq. (4.22), that the generalized formalism allows the existence of such state (or states) of the motion, where $\omega_g = 0$ despite $H_0 \neq 0$ (the conventional formalism allows to be $\omega_0 = |2H_0|^{3/2}/D^{1/2} = 0$ only for $H_0 = 0$). This is a state where $B(H_0, A_0) = -1$. Therefore, *such state or states would not emit the electromagnetic radiation, would not lose energy for the radiation, and would thus constitute stable states of the classical atom.*

It important to point out that the physics behind such classical non-radiating states is a *new kind of time-dilation* expressed by Eq. (4.21). This is a *non-Einsteinian time-dilation*. The closer the energy of the system to the energy of the classical non-radiating state, the more dilates the time. At the classical non-radiating state, the time gets dilated infinitely, so that the frequency $\omega_{\mathrm{g}}$ in Eq. (4.22) vanishes and so does the radiation.

The function $B(H_0, A_0)$ can be determined based on the following experimental fact. Experiments show that a weakly bounded hydrogenic atom ($H_0 \to 0$) emits the radiation primarily at some *frequency tending to a finite non-zero limit* as $H_0 \to 0$. (We use only the existence of such limiting frequency of the radiation without using the experimental value for this limiting frequency.) Therefore, as $H_0 \to 0$, we should have from Eq. (4.22):

$$\omega_g \approx |B(H_0, A_0)||2H_0|^{3/2}/D^{1/2} = \text{const} \equiv Q, \qquad (4.23)$$

so that

$$B(H_0, A_0) = QD^{1/2}/|2H_0|^{3/2}. \qquad (4.24)$$

Now we consider the energy $H_0 = H_{st}$ such that at this (yet unknown) energy there is no radiation because the frequency $\omega_g$ becomes zero. In accordance to above, this occurs at

$$B(H_{st}, A_{st}) = -1. \qquad (4.25)$$

Combining Eqs. (4.24) and (4.25), we find $Q = -|2H_{st}|^{3/2}/D^{1/2}$, so that

$$B(H_0, A_0) = -|H_{st}/H_0|^{3/2}. \qquad (4.26)$$

Thus, in the classical process of the radiation from a weakly bounded state, the frequency of the radiation is

$$\omega_g(H_0) = (8/D)^{1/2}[|H_{st}|^{3/2} - |H_0|^{3/2}]. \qquad (4.27)$$

Until the radiation vanishes, its frequency varies from $\omega_g^{init} \approx (8/D)^{1/2}|H_{st}|^{3/2}$ to $\omega_g^{fin} = 0$. Therefore, the *average* frequency in the process of the radiation is

$$\langle \omega_g \rangle \approx (\omega_g^{init} + \omega_g^{fin})/2 \approx (2/D)^{1/2}|H_{st}|^{3/2}. \qquad (4.28)$$

In accordance to the Plank's hypothesis (later incorporated by Einstein into his photoelectric law — see, e.g., [4.6]), the change of the energy $\Delta E = |H_{st}| - |H_0| \approx |H_{st}|$ should be equal to $\hbar\langle\omega_g\rangle$:

$$|H_{st}| \approx \hbar\langle\omega_g\rangle \approx \hbar(2/D)^{1/2}|H_{st}|^{3/2}. \qquad (4.29)$$

Thus we find the absolute value of the energy of the radiationless stable state to be

$$|H_{st}| = D/(2\hbar^2) = \mu Z^2 e^4/(2\hbar^2). \qquad (4.30)$$

The value we obtained coincides exactly with the experimental result for the ionization potential of hydrogenic atoms. The function $B(H_0)$ is now completely defined by Eqs. (4.26) and (4.30).

It should be emphasized that in the above derivation we employed two fundamental experimental facts, but *did not "forcefully" quantize any physical quantity describing the atom*. In particular, this amounts to *classically deriving the Bohr's postulate* on the quantization of the angular momentum (instead of postulating it).

The above formalism allows deducing not only the lowest non-radiating stable state, but actually all other non-radiating stable states as well. For doing this, we should recognize that the electron, while executing the elliptical orbit around the nucleus, emits the classical electromagnetic radiation not only at the primary frequency of the revolution $\omega_g$, but also at its harmonics (see, e.g., [4.7]): $\omega_r = k\omega_g$, $k = 1, 2, 3, \ldots$. Therefore, instead of Eq. (4.22), we should actually have:

$$\omega_g = k|1 + B_k(H_0, A_0)||2H_0|^{3/2}/D^{1/2}. \tag{4.31}$$

As $H_0 \to 0$, we get from Eq. (4.31):

$$\omega_g \approx k|B_k(H_0, A_0)||2H_0|^{3/2}/D^{1/2} = \text{const} \equiv Q_k, \tag{4.32}$$

so that

$$B_k(H_0, A_0) = Q_k D^{1/2}/|2H_0|^{3/2}. \tag{4.33}$$

Now we consider the energy $H_0 = H_{st}(k)$ such that at this (yet unknown) energy there is no radiation because the frequency $\omega_r$ becomes zero. This occurs at

$$B_k(H_{st}(k), A_{st}(k)) = -1. \tag{4.34}$$

Combining Eqs. (4.33) and (4.34), we find $Q_k = -|2H_{st}(k)|^{3/2}/D^{1/2}$, so that

$$B_k(H_0, A_0) = -|H_{st}(k)/H_0|^{3/2}. \tag{4.35}$$

In the classical process of the radiation from a weakly bounded state, the frequency of the radiation is

$$\omega_g(H_0) = k(8/D)^{1/2}[|H_{st}(k)|^{3/2} - |H_0|^{3/2}]. \tag{4.36}$$

Until the radiation vanishes, its frequency varies from $\omega_r^{\text{init}} \approx k(8/D)^{1/2}|H_{st}(k)|^{3/2}$ to $\omega_r^{\text{fin}} = 0$. Therefore, the *average* frequency in the process of the radiation is

$$\langle \omega_g \rangle \approx (\omega_g^{\text{init}} + \omega_g^{\text{fin}})/2 \approx k(2/D)^{1/2}|H_{\text{st}}(k)|^{3/2}. \qquad (4.37)$$

In accordance to the Plank's hypothesis (later incorporated by Einstein into his photoelectric law — see, e.g., [4.6]), the change of the energy $\Delta E = |H_{\text{st}}(k)| - |H_0| \approx |H_{\text{st}}(k)|$ should be equal to $\hbar\langle\omega_g\rangle$:

$$|H_{\text{st}}(k)| \approx \hbar\langle\omega_r\rangle \approx \hbar k(2/D)^{1/2}|H_{\text{st}}(k)|^{3/2}. \qquad (4.38)$$

Thus we find the absolute value of the energy of the non-radiating stable states to be

$$|H_{\text{st}}(k)| = D/(2\hbar^2 k^2) = \mu Z^2 e^4/(2\hbar^2 k^2) \equiv -E_k, k = 1, 2, 3, \ldots. \qquad (4.39)$$

The value we obtained coincides exactly with the experimental result for the binding energy of stationary states of hydrogenic atoms. The integer $k$ in Eq. (4.39) is not a "mysterious" quantum number, but rather a number of the classical harmonic, at which the classical electromagnetic radiation is emitted by the atomic electron. The choice of

$$B_k(\text{H}_0) = -|E_k/H_0|^{3/2} \qquad (4.40)$$

determines a Hamiltonian $H_g^{(k)}$ (see (4.18)), which classically reproduces a hydrogenic atom with a single non-radiating energy level $k$.

It is also feasible to classically reproduce a hydrogenic atom having $N$ non-radiating energy levels ($k = 1, 2, \ldots, N$) — by choosing

$$B(H_0) = -\sum_{k=1}^{N} \theta(E_{k+1} - H_0)\theta(H_0 - E_k)|E_k/H_0|^{3/2}, \qquad (4.41)$$

where $\theta(\text{x})$ is the step-function: $\theta(x) = 1$ for $x > 0$, $\theta(x) = 0$ for $x \leq 0$.

Let us discuss a classical non-radiating stable state in more detail. In such a state we have $dr/dt = dp/dt = 0$, so that $r(t) = r_0$ and $p(t) = p_0$, where $r_0$ and $p_0$ are some vector constants. Thus, *the electron is motionless, but its momentum differs from zero.* This is not surprising: the momentum $\mathbf{p}$ is a more complex physical quantity than the velocity $\mathbf{v} \equiv dr/dt$. For example, for a charge in an electromagnetic field characterized by a vector-potential $\mathbf{A}$, it is also possible to have $\mathbf{v} = [\mathbf{p} - e\mathbf{A}/(mc)]/m = 0$ while $\mathbf{p} = e\mathbf{A}/(mc) \neq 0$ [4.7].

We developed a purely classical formalism for atomic and molecular systems, including the electronic motion. We employed the Dirac's GHD designed for systems having constraints and applied it to atomic and molecular physics by choosing integrals of the motion as constraints. We considered in detail the application of our formalism to hydrogenic atoms.

It might seem that hydrogenic atoms would not constitute a "typical" example from atomic and molecular physics because it belongs to quantum systems, which have a higher symmetry than their geometrical symmetry and therefore possess "additional" integrals of the motion (sometimes called an "accidental" degeneracy). Examples other than an isolated hydrogenic atom are three-dimensional isotropic harmonic oscillator, a hydrogen atom in a uniform electric field, an electron in the field of two Coulomb centers. Further, it is usually considered that such a situation is exceptional and that the overwhelming majority of atoms and molecules possess only integrals of the motion dictated by the geometrical symmetry (e.g., the angular momentum vector for the spherical symmetry or its projection for the axial symmetry).

However, the latter is a misconception. In the reality, it turns out that, for instance, *all classical* spherically symmetric potentials have the $O_4$ and $SU_3$ symmetries [4.8–4.10]. Therefore, all such potentials possess an additional vector integral of the motion (an analog of the Runge–Lenz vector), which is an element of the Lie algebra and to which *no quantal operator can correspond* [4.8–4.10]. It is significant that the Poisson brackets of this classical vector integral of the motion with other integrals of the motion are *the same as for the Runge–Lenz vector.*

The most important is the following non-trivial fact. *All classical systems of N degrees of freedom have the $O_{N+1}$ and $SU_N$ symmetries regardless of the functional form of the Hamiltonian* [4.10]. In particular, *all systems of 3 degrees of freedom* (not only spherically-symmetric ones) possess an additional vector integral of the motion.

We emphasize again that for the overwhelming majority of the microscopic systems, the additional ("non-geometrical") integrals of the motions exist only classically, but not quantally (few exceptions, such as a hydrogen atom etc. had been mentioned above). Thus, *classical mechanics has an intrinsic, built-in advantage over quantum mechanics* in this regard, but this advantage was never employed to the fullest.

The most immediate application of our formalism should be to chemical physics. Indeed, it is well-known that an accurate quantal description of polyatomic molecules is, in general, virtually impossible. Therefore, in chemical physics a number of approaches were proposed for incorporating electronic degrees of freedom into a classical theory, such as various versions of the surface hopping model [4.11–4.15], "rigorous" semiclassical methods [4.16–4.18], and classical path models where classical nuclear motion is coupled to time-dependent electronic motion [4.19–4.22]. Out of these models, one of the most popular seems to be the classical electron analog (CEA) model by Meyer and Miller [4.21]. The basic idea of the CEA is to replace a set of electronic states by a set of classical action-angle variables, the action variables typically being populations of the corresponding electronic states. The CEA model was successfully applied to a number of chemical processes — see, e.g., paper [4.23], containing a further development of the CEA model, and references therein.

However, with all due respect to the ingenuity of the authors of the CEA model, our purely classical formalism has significant advantages over it. First, the CEA model requires (as an input) matrix elements of the interaction coupling electronic states; thus, the CEA model requires the knowledge of quantal wave functions (as well as quantal eigenvalues). In distinction, our formalism does not require as an input the knowledge of the quantal wave functions (and matrix elements are calculated on the basis of these wave functions).

Second, the CEA model does not address one of the most fundamental questions: *what is a purely classical (not semiclassical) distinction of trajectories, which quantally correspond to stationary electronic states?* In other words, the CEA model implicitly adopted the Bohr's quantization postulate, thus making the model, in fact, semiclassical rather than purely classical. In distinction, in our purely classical formalism we actually derived the Bohr's quantization postulate instead of postulating it.

## 4.2. Extending GHD to the Motion in a Modified Coulomb Potential

To extend the GDH to other atomic and molecular systems, one needs explicit expressions for all integrals of the motion. Here we derive the explicit form of the additional integral of the motion — the unit Runge–Lenz vector — for a modified Coulomb potential. Namely, we consider a Coulomb or Kepler potential (proportional to $1/r$) modified by adding a term proportional to $1/r^2$ — called hereafter the *binomial potential*. This potential has interesting applications. Examples are as follows.

1. *Pionic and kaonic atoms.* Since pions and kaons are spinless particles, pionic and kaonic atoms are described by the quantal (relativistic) Klein–Gordon equation. The radial Klein–Gordon equation for the problem of a pionic or kaonic atom is given by (see, e.g., [4.24–4.27]):

$$\frac{d^2 R}{d\rho^2} + \frac{2}{\rho}\frac{dR}{d\rho} + \left[\frac{\lambda}{\rho} - \frac{1}{4} - \frac{l(l+1) - (Z\alpha)^2}{\rho^2}\right] R = 0, \qquad (4.42)$$

where $Z$ is the nuclear charge and $\alpha = \frac{e^2}{ch} \cong \frac{1}{137}$ is the fine structure constant. Other notations in Eq. (4.42) are as follows:

$$\lambda = Z\alpha E/(M^2 c^4 - E^2)^{1/2}, \quad \rho = \beta r,$$
$$\beta = 2(M^2 c^4 - E^2)^{1/2}/(\hbar c), \qquad (4.43)$$

where $E$ is the energy. The Klein–Gordon equation (4.42) for the Coulomb potential is equivalent to the radial Schrödinger equation with a potential $U$ and an energy $W$, such that

$$U/(4W) = \lambda/\rho + (Z\alpha)^2/\rho^2, \quad (W < 0), \qquad (4.44)$$

which is the binomial potential.

2. *Precessions of planetary orbits.* In his seminal paper, *Die Grund-lange der allgemeinen Relativitästhoerie* [4.28], Einstein showed that general relativistic effects perturb the Kepler potential by an additive term proportional to $1/r^2$ and used it to calculate the precession of Mercury's orbit around the sun. This fundamental result can be also found, e.g., in [4.7, 4.29, 4.30].

3. *Nanoplasmas.* Karnakov *et al.* [4.31] derived the spectrum and expressions for the intensity of dipole radiation for a classical non-relativistic particle executing non-periodic motion. The potential in which the particles under consideration move is of the form $U(r) = -\alpha/r + \beta/r^2$. Karnakov *et al.* [4.31] applied their results to the description of the radiation and the absorption of a clas-sical collisionless electron plasma in nanoparticles irradiated by an intense laser field. They also obtained the rate of collisionless absorption of electromagnetic wave energy in equilibrium isotropic nanoplasma.

The binomial potential in classical mechanics possesses an addi-tional integral of the motion in addition to the energy and the angu-lar momentum vector. This is a particular case of a general result by Fradkin [4.9]. He showed that all classical dynamical problems of both the relativistic and non-relativistic type, dealing with a central potential, necessarily possess $O(4)$ and $SU(3)$ symmetries, and thus should have integrals of the motion additional to the energy and the angular momentum vector. He introduced a unit Runge–Lenz vector, which is a generalization of the Runge–Lenz vector well-known for the Coulomb/Kepler potential (see, e.g., [4.5]).

Holas and March [4.32] provided a further development of the unit Runge vector. In distinction to Fradkin, they focused on the construction and time dependence of the vector itself rather than on the dynamical symmetries of central potentials or the algebras satisfied by the unit Runge vector.

Here we use the results of Holas and March [4.32] to derive the explicit form of the unit Runge–Lenz vector for the binomial potential.

Fradkin [4.9] has shown that all classical dynamical problems of both the relativistic and non-relativistic type, dealing with a central potential, necessarily possess $O(4)$ and $SU(3)$ symmetries, as mentioned above. This led him to a generalization of the Runge–Lenz vector. Here we will briefly present his results relating to the generalization of the Runge–Lenz vector and the construction of the elements of the Lie algebra of $O(4)$ and $SU(3)$ in terms of canonical variables.

In the non-relativistic classical Coulomb/Kepler problem, the force on the affected particle is an inverse square force given by

$$\dot{\mathbf{p}} = -\frac{\lambda}{r^2}\hat{r}; \quad \mathbf{p} = m\dot{\mathbf{r}}, \quad \hat{r} = \frac{\mathbf{r}}{r}, \tag{4.45}$$

where the overdot denotes the total derivative with respect to time. In the Coulomb/Kepler problem, the Hamiltonian and the angular momentum vector $\mathbf{L}$ are the conserved quantities. There also exists another conserved vector quantity, namely the Laplace–Runge–Lenz vector, or simply the Runge–Lenz vector defined as

$$\mathbf{A} = (-2mE)^{-\frac{1}{2}}(\mathbf{p} \times \mathbf{L} - \lambda m\hat{r}). \tag{4.46}$$

This vector lies in the plane of the orbit and points from the nucleus to the perihelion of the orbit, as shown in Fig. 4.1.

Fradkin found, via the standard Poisson bracket formalism, that for the classical Coulomb/Kepler problem, and indeed for all central potential problems, $\mathbf{A}$, $\mathbf{L}$, and $H$ satisfy the following closed Lie

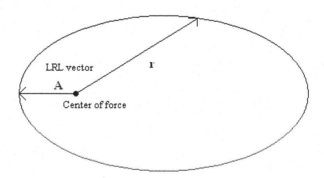

Fig. 4.1. Direction of the Laplace–Runge–Lenz vector $\mathbf{A}$ within the elliptic trajectory, corresponding to the motion in a Coulomb or Kepler potential.

algebra:

$$
\begin{aligned}
&[A_i, H] = [L_i, H] = 0; \\
&[L_i, L_j] = \varepsilon_{ijk} L_k; \\
&[L_i, A_j] = \varepsilon_{ijk} A_k; \\
&[A_i, A_j] = \varepsilon_{ijk} L_k.
\end{aligned}
\tag{4.47}
$$

It is seen that the Lie algebra given above is isomorphic to that of the generator of the $O(4)$ symmetry group, which is the group of orthogonal transformations representing rotations in four dimensions. Fradkin also concluded that if the existence of the Runge–Lenz vector is simply to ensure that the plane of the motion is conserved, then it should always be possible to find a vector analogous to the Runge–Lenz vector *for all central potentials*.

Fradkin proposed a generalization for the Runge–Lenz vector by choosing $\hat{r}$, $\hat{L}$, and $\hat{r} \times \hat{L}$ as a mutually orthogonal triad of unit vectors. This unit Runge vector is

$$
\hat{k} = (\hat{k} \cdot \hat{r})\hat{r} + (\hat{k} \cdot \hat{L})\hat{L} + (\hat{k} \cdot \hat{r} \times \hat{L})\hat{r} \times \hat{L},
\tag{4.48}
$$

but since the unit Runge vector is in the plane of the orbit and the angular momentum vector is perpendicular to the plane of motion, then the second term is identically zero ($\hat{k} \perp \mathbf{L}$). $\hat{k}$ may be chosen to be the direction from which the azimuthal angle $\theta$ is measured (with the positive sign given by a right-handed rotation about $\hat{L}$), then one has

$$
\hat{r} \cdot \hat{k} = \cos\theta \quad \text{and} \quad \hat{k} \cdot \hat{r} \times \hat{L} = \sin\theta,
\tag{4.49}
$$

so that

$$
\hat{k} = (\cos\theta)\hat{r} + (\sin\theta)\hat{r} \times \hat{L}.
\tag{4.50}
$$

Defining $u = 1/r$, one can write the following differential equation for $u$ and the azimuthal angle $\theta$ in terms of the energy $E$, potential

$V$ and angular momentum $L$:

$$\left(\frac{du}{d\theta}\right)^2 = \left(\frac{2m}{L^2}\right)(E - V) - u^2. \tag{4.51}$$

At this point we note the following relations and definition:

$$\cos\theta = f(u, L^2, E),$$

$$\sin\theta = \left(\frac{\partial f}{\partial u}\right)\frac{(\hat{r} \cdot \mathbf{p})}{L}. \tag{4.52}$$

Further, after putting $V = -\lambda u$ for the potential of the Coulomb/Kepler problem, the equation of the orbit becomes:

$$f = \cos\theta = [2mEL^2 + (\lambda m)^2]^{-1/2}(L^2 u - \lambda m). \tag{4.53}$$

The unit Runge vector can be expressed as:

$$\hat{k} = \left[f - u\frac{\partial f}{\partial u}\right]\hat{r} + L^{-2}\frac{\partial f}{\partial u}\mathbf{p} \times \mathbf{L}. \tag{4.54}$$

Its Poisson bracket with the Hamiltonian function vanishes. Poisson brackets between the components of the unit Runge vector also vanish. A complete set of Poisson brackets involving the unit Runge vector is the following:

$$[\hat{k}_i, H] = 0; \quad [\hat{k}_i, \hat{k}_j] = 0; \quad [L_i, \hat{k}_j] = \varepsilon_{ijk}\hat{k}_k; \quad \text{for } i, j, k = 1, 2, 3. \tag{4.55}$$

Holas and March [4.32] provided a further development of the unit Runge vector. They focused on the construction and the time dependence of the vector itself rather than on the dynamical symmetries of central potentials or the algebras satisfied by the unit Runge vector.

Holas and March used the relation

$$\mathbf{p} \times \mathbf{L} = \frac{\mathbf{r}L^2}{r^2} - \frac{(\mathbf{p} \cdot \mathbf{r})}{r^2 L}\mathbf{L} \times \mathbf{r} \tag{4.56}$$

to rewrite the unit Runge vector as

$$\hat{k} = f\hat{r} - \frac{(\mathbf{p} \cdot \mathbf{r})}{Lr}\frac{\partial f}{\partial u}\hat{L} \times \hat{r}, \tag{4.57}$$

where the function $f$ is specified in the next section. This is the implicit form of the unit Runge vector, from which we shall start deriving the explicit form for the binomial potential.

The classical Hamilton function for the binomial potential under consideration can be written in the form:

$$H = \frac{p^2}{2\mu} - \frac{C}{r} + \frac{\Lambda}{2\mu r^2}. \qquad (4.58)$$

Here $\mu$ is the reduced mass; the constant $C$ is equal to either $Ze^2$ for the Coulomb potential ($Z$ is the nuclear charge) or to $Gm_1m_2$ for the Kepler potential ($G$ is the gravitational constant).

The function $f$ is given by

$$f = \cos\theta; \theta = \frac{L}{\mu} \int_{u_0}^{u} \left[ \frac{2}{\mu}\left(E - V\left(\frac{1}{u'}\right)\right) - \left(\frac{L_{eff}u'}{\mu}\right)^2 \right]^{-1/2} du',$$
$$(4.59)$$

where $V(r)$ is the Coulomb/Kepler part of the binomial potential and

$$L_{eff}^2 = L^2 - \Lambda \qquad (4.60)$$

is the effective angular momentum. The second term in Eq. (4.60) corresponds to the presence of the term proportional to $1/r^2$ in the binomial potential. The integral in Eq. (4.59), upon the substitution of the Coulomb/Kepler potential, can be rewritten as:

$$\int_{u_0}^{u} \left[ \frac{2}{\mu}\left(E - V\left(\frac{1}{u'}\right)\right) - \left(\frac{L_{eff}u'}{\mu}\right)^2 \right]^{-1/2} du'$$
$$= \int_{u_0}^{u} \frac{du'}{\sqrt{\left(\frac{2}{\mu}\right)(E + Ze^2u') - \left(\frac{Lu'}{\mu}\right)^2}}. \qquad (4.61)$$

After the substitutions

$$u_1 = \frac{Ze^2\mu}{L^2}\left(1 + \sqrt{\frac{2EL^2}{Z^2e^4\mu}}\right),$$

$$u_2 = \frac{Ze^2\mu}{L^2}\left(1 - \sqrt{\frac{2EL^2}{Z^2e^4\mu}}\right), \qquad (4.62)$$

the left-hand side of Eq. (4.61), in the indefinite form of the integral, becomes (after some simplifications):

$$\int \frac{du'}{\sqrt{(u_1 - u)(u - u_2)}} = \tan^{-1}\left(\frac{u - \frac{u_1 + u_2}{2}}{\sqrt{(u_1 - u)(u - u_2)}}\right). \qquad (4.63)$$

It is convenient to define

$$u_3 = \frac{u_1 + u_2}{2} = \frac{\mu Z e^2}{L^2}, \qquad (4.64)$$

so that Eq. (4.63) reduces to:

$$\tan^{-1}\left(\frac{u - \frac{u_1 + u_2}{2}}{\sqrt{(u_1 - u)(u - u_2)}}\right) = \tan^{-1}\left(\frac{u - u_3}{\sqrt{(u_1 - u)(u - u_2)}}\right). \qquad (4.65)$$

Entering the limits of integration yields:

$$f = \cos\left(\tan^{-1}\left(\frac{u - u_3}{\sqrt{(u_1 - u)(u - u_2)}}\right)\right.$$
$$\left. - \tan^{-1}\left(\frac{u_0 - u_3}{\sqrt{(u_1 - u_0)(u_0 - u_2)}}\right)\right). \qquad (4.66)$$

It is convenient to define:

$$g = g(u) \equiv \frac{u - u_3}{\sqrt{(u_1 - u)(u - u_2)}}; \quad g(u_0) \equiv g_0. \qquad (4.67)$$

Using the identity

$$\cos(\tan^{-1}(A) - \tan^{-1}(B)) = \frac{1 + AB}{\sqrt{1 + A^2 + B^2 + A^2 B^2}}, \qquad (4.68)$$

we can then write

$$f = \frac{1 + g g_0}{\sqrt{1 + g^2 + g_0^2 + g^2 g_0^2}}. \qquad (4.69)$$

Consequently, the partial derivative in the unit Runge vector becomes:

$$\frac{\partial f}{\partial u} = \left[\frac{g_0}{\sqrt{1 + g^2 + g_0^2 + g^2 g_0^2}} - \frac{(1 + g g_0)(g + g g_0^2)}{(1 + g^2 + g_0^2 + g^2 g_0^2)^{3/2}}\right]\left(\frac{\partial g}{\partial u}\right), \qquad (4.70)$$

where

$$\frac{\partial g}{\partial u} = \frac{1}{\sqrt{(u_1 - u)(u - u_2)}} - \frac{1}{2}\frac{\left(u - \frac{u_1 + u_2}{2}\right)(-2u + u_1 + u_2)}{((u_1 - u)(u - u_2))^{3/2}}. \qquad (4.71)$$

Now we use the definitions (4.67) and (4.69) to rewrite Eq. (4.71) and substitute it into Eq. (4.70) to get the following compact form:

$$\frac{\partial f}{\partial u} = \left( \frac{g_0 f}{1 + gg_0} - \frac{g(1 + g_0^2)}{(1 + gg_0)^2} f^3 \right) \left( \frac{g - g^3}{u - u_3} \right). \tag{4.72}$$

Here the term in the second set of parenthesis is the simplification of $\partial g / \partial u$. We thus arrive at:

$$\hat{k} = \frac{1 + gg_0}{\sqrt{1 + g^2 + g_0^2 + g^2 g_0^2}} \hat{r} - \frac{\mathbf{p} \cdot \mathbf{r}}{Lr} \left( \frac{g_0 f}{1 + gg_0} - \frac{g(1 + g_0^2)}{(1 + g \, g_0)^2} f^3 \right)$$
$$\times \left( \frac{g - g^3}{u - u_3} \right) \hat{L} \times \hat{r}. \tag{4.73}$$

Equation (4.73), with $f$ defined by Eq. (4.69), represents the explicit form of the unit Runge–Lenz vector for the binomial potential. This is a general result valid for any value of $\Lambda$.

In some applications, the second term in the binomial potential can be considered as a small perturbation, so that $\Lambda \ll L^2$. Further simplifications for this particular case can be found in paper [4.33].

Our results can be immediately used to extend the GHD to the physical systems described by the binomial potential. The angular momentum vector and the unit Runge vector are constants of the motion for a centrally symmetric potential and consequently have vanishing Poisson brackets with the Hamiltonian for the system and are thus suitable constraints for the application of GHD. The initial representation of the generalized Hamiltonian for this system is:

$$H_g = \frac{p^2}{2\mu} - \frac{Ze^2}{r} + \frac{\Lambda}{2\mu r^2} + \mathbf{u} \cdot (\mathbf{L} - \mathbf{L}_0) + \mathbf{w} \cdot (\hat{k} - \hat{k}_0), \tag{4.74}$$

where $\Lambda$ is the strength of the binomial potential, $Ze$ is the nuclear charge, $e$ is the electron charge, $\mu$ is the reduced mass, $\mathbf{u}$ and $\mathbf{w}$ are yet unknown constant vectors (to be determined later) of the GHD formalism. In Eq. (4.74), $\mathbf{L}_0$ and $\hat{k}_0$ are the values of the angular momentum and unit Runge vector in a particular state of the motion so that in those states

$$\mathbf{L} \approx \mathbf{L}_0 \qquad (4.75)$$

and

$$\hat{k} \approx \hat{k}_0. \qquad (4.76)$$

We define the following quantities:

$$H_0 = \frac{p^2}{2\mu} - \frac{Ze^2}{r},$$

$$H_B = H_0 + \frac{\Lambda}{2\mu r^2}, \qquad (4.77)$$

where the subscript $B$ stands for binomial. The consistency conditions for this system are:

$$[\mathbf{L}, H_g] \approx 0,$$

$$[\hat{k}, H_g] \approx 0. \qquad (4.78)$$

The next step is to calculate the Poisson brackets given in Eq. (4.78) to arrive at a functional form of the consistency conditions and thus solve for the unknown vector coefficients $\mathbf{u}$ and $\mathbf{w}$. We begin with the angular momentum bracket:

$$[L_i, H_g] = [L_i, H_0] + \left[L_i, \frac{\Lambda}{2\mu r^2}\right] + u_j[L_i, (L_j - L_{0_j})]$$

$$+ w_j[L_i, (\hat{k}_j - \hat{k}_{0j})]. \qquad (4.79)$$

Clearly the first two terms vanish since the angular momentum is conserved in any centrally symmetric potential in the absence of external forces. So we obtain:

$$[\mathbf{L}, H_g] = \mathbf{u} \times \mathbf{L} + \mathbf{w} \times \hat{k} \approx 0. \qquad (4.80)$$

We now proceed to the calculation of the time derivative of the unit Runge vector via the Poisson bracket:

$$[\hat{k}, H_g] = [\hat{k}_i, H_0] + \left[\hat{k}_i, \frac{\Lambda}{2\mu r^2}\right] + u_j[\hat{k}_i, (L_j - L_{0_i})]$$

$$+ w_j[\hat{k}_i, (\hat{k}_j - \hat{k}_{0j})] \approx 0. \qquad (4.81)$$

The following result is obtained:

$$[\hat{k}, H_g] = \mathbf{u} \times \hat{k} \approx 0. \tag{4.82}$$

A well-known relation between $A$, $L$, and $H_0$ is:

$$A = \sqrt{1 + \frac{2H_0 L^2}{\mu Z e^2}}. \tag{4.83}$$

We seek the unknown vector coefficients in the following form:

$$\begin{aligned}
\mathbf{u} &= a_1 \hat{k}_0 + a_2 \mathbf{L}_0 + a_3 \hat{k}_0 \times \mathbf{L}_0, \\
\mathbf{w} &= b_1 \hat{k}_0 + b_2 \mathbf{L}_0 + b_3 \hat{k}_0 \times \mathbf{L}_0.
\end{aligned} \tag{4.84}$$

Substituting Eq. (4.84) into Eq. (4.80), we get:

$$a_1 \hat{k}_0 \times \mathbf{L}_0 + a_3 (\hat{k}_0 \times \mathbf{L}_0) \times \mathbf{L}_0 - b_2 \hat{k}_0 \times \mathbf{L}_0 + b_3 (\hat{k}_0 \times \mathbf{L}_0) \times \hat{k}_0 \approx 0. \tag{4.85}$$

For this expression to vanish, combined with the requirement for Eq. (4.82) to be satisfied, we conclude that

$$a_1 = b_2 \quad \text{and} \quad a_2 = a_3 = b_3 = 0. \tag{4.86}$$

So, finally we obtain:

$$\begin{aligned}
\mathbf{u} &= a_1 \hat{k}_0, \\
\mathbf{w} &= b_1 \hat{k}_0 + a_1 \mathbf{L}_0.
\end{aligned} \tag{4.87}$$

Thus, the generalized classical Hamiltonian for the binomial potential is defined by Eq. (4.74) with the vector coefficients $\mathbf{u}$ and $\mathbf{w}$ given by Eq. (4.87). It contains only two unknown coefficients $a_1$ and $b_1$ instead of the initially six unknown components of $\mathbf{u}$ and $\mathbf{w}$. Further details can be found in paper [4.33].

# Chapter 5

# Underestimated Role of the Singular Spin–Spin Interaction in the Binding Energy of Two-Electron Atoms/Ions

Indications of a significant role of correlation interactions of electrons have been obtained in studies of collisions in simple atomic systems ($H^+$, $H^-$, and $H^+$) [5.1–5.5]. Such correlations are usually attributed to the Coulomb interaction complicated by symmetrization of wavefunctions. However, this symmetrization is a consequence of the spin magnetic moments of electrons. The contribution of the spin–spin interaction to the expression for the binding energy appears with a small parameter $\alpha^2 = e^4/\hbar^2 c^2$, which suggests that it is manifested only in calculation of the fine structure of spectra [5.6–5.8]. The attempt at taking into account the contribution from the spin–spin interaction directly was made as early as in the 1920s [5.6, 5.9, 5.10] in determining excited energy levels of singlet and triplet structures. In contrast to the spin–orbit interaction, the spin–spin interaction depends not on the distance between the electrons and the nucleus, but only on the relative distance between the electrons. However, the relative distance $r_{12}$ in [5.6, 5.9, 5.10] was replaced by the distance from the nucleus and an excited electron because this electron is always separated from the nucleus by a much larger distance than an unexcited electron. As a result, the role of the singularity $|\mathbf{r}_{ij}|^{-3}$ in the spin–spin interaction in the spatial region in which the distance $r_{12}$ between the electrons is smaller than the Compton wavelength $\lambda_C = \hbar/mc$ was disregarded. At that time, the interest of researches was mainly concentrated on the description of the fine structure of triplet energy levels. Since the allowance for the spin–spin interaction could only lead to a displacement of terms as a whole, interest in a more detailed account of the spin–spin interaction was lost.

This study is devoted to theoretical analysis of the contribution of the spin–spin interaction to the correlation energy for two-electron atoms/ions. Usually, the correlation energy is treated as the contribution to the binding energy from the corrections to the wavefunction, which exceed the limits of the Hartree–Fock approximation, but are not connected directly with any other interaction except the Coulomb interaction [5.11]. However, analysis of the Hamiltonian of a two-electron atom [5.6–5.8] clearly shows that correlations may also appear due to the spin–spin interaction between the electrons, which is weak but has a strong singularity; this interaction is determined by the relative distance between the electrons and is independent of their separation from the center of mass of the system. Depending on the type of singularity, the small parameter determining the intensity of this interaction for moderate distances between the particles cannot guarantee the smallness of this contribution when the relative distance tends to zero.

The purpose of this work is to construct an algorithm taking into account the role of this type of correlations in the formation of the binding energy of two-electron systems in their ground states, where it is impossible for the electron-nucleus separation of one electron to significantly differ from the electron–nucleus separation of the other electron. The algorithm is constructed for a two-electron system in the non-relativistic approximation with the spin–spin interaction Hamiltonian in the form used for calculating the fine structure of the spectra [5.8, 5.12–5.14]. An important result is the establishment of the compensation of the smallness $\sim \alpha^2$ by the singularity of the interelectron interaction proportional to $|\mathbf{r}_{ij}|^{-3}$. As a result, the contribution of the spin–spin attraction to the binding energy of the singlet ground state, under the condition $r_{12} \leq r_{0Z} = Ze^2/(mc^2)$ becomes comparable by magnitude and opposite by sign to the contribution from the Coulomb repulsion. The scale $r_{0Z}$ of the correlated motion is a natural physical limit restricting the range of the independent motion of electrons having a singular interaction between their spins in the configuration space. Such a spatially correlated motion of electrons can be taken into account by expanding the basis of the ground-state functions via the introduction into this basis of a "paired" hydrogen-like state with principal quantum number $n = 2$ for an electron with charge $2e$ and mass $2m_e$ in the field of a nucleus with charge $Ze$. In the zeroth approximation, the sum of the binding energies of the two non-interacting $1s$ electrons coincides with the energy of the "paired" state with $n = 2$. Such a possibility of extending the basis is ensured by the increase in the dimension of the configuration space and the change in its structure because the spin moments of electrons and their magnetic interaction are taken into account.

The relative motion of the two electrons and the motion of their center of mass can be separated by disregarding terms $\sim \alpha^2$. The wave function of the relative motion of the paired electrons can be represented in the form

$$\varphi(r_{12}) = \lambda r_{12}^{-1/4}/(r_{12}^2 + r_{0Z}^2)^2, \tag{5.1}$$

where $\lambda$ is the normalization coefficient. After calculating the normalization integral, one obtains:

$$\lambda = 2^{15/4} r_{0Z}^{15/2}/(77\pi)^{1/2}. \tag{5.2}$$

The derivation of Eq. (5.1) was based in part on the well-known momentum representation of the wave function of $1s$ electrons in the Coulomb field of the nucleus with the characteristic momentum $mv_{12}$, where $v_{12} = |r_{12}|mc^2/\hbar$ is the velocity of the drift motion of the electrons around the center of mass. The drift motion is due to the fact that in a stationary state, the Coulomb repulsion of the electrons is balanced by the Lorentz force resulting from the magnetic field $B$ produced by the spin magnetic moment $\mu$ of the neighboring electron: $B \sim |\mu/r_{12}^3|$.

The increase in the number of basis functions necessitates the introduction of various spatial coordinates depending on the inter-electron separation $r_{12}$ and, as a consequence, the use of the density matrix formalism. In the range of $r_{12} > \lambda_C$, the contribution from the spin-spin interaction is found to be on the order of $\alpha^2$, and the approximation of independent electrons is valid. In the case of a *singlet* state, in the range of $r_{12} > r_{0Z}$, a finite motion appears, and the motion of the center of mass of the electrons is described by a hydrogen-like wavefunction with the principal quantum $n = 2$. The evolution of the spatial scale of the paired state proceeds from $(r_{12})_{\min} \sim \lambda_C$ at the moment of the formation of the pair to $(r_{12})_{\min} \sim r_{0Z}$ (when the distance between the electron center of mass and the nucleus increases to approximately $a_{0Z} = a_0/Z$, where $a_0$ is the Bohr radius).

The fact that the finite motion appears for a singlet state, but not for a triplet state is a simple consequence of the well-known expression for the spin–spin interaction

$$V_{SS} \sim \alpha \mathbf{S}_1 \cdot \mathbf{S}_2. \tag{5.3}$$

The product $\mathbf{S}_1 \cdot \mathbf{S}_2$ is negative (equal to $-3/4$) for antiparallel spins, what effectively corresponds to an attraction. However, this product is positive (equal to $1/4$) for parallel spins, what effectively corresponds to a repulsion.

Employing the above concepts, a good agreement with experimental values of the ionization potential is obtained for a wide range of two-electron atoms/ions without resorting to variational procedures, but using only hydrogen-like wavefunctions and taking into account the singular nature of the spin–spin interaction between the electrons correctly. Essentially, two expressions for the binding energy of two-electron systems are obtained. The first one is:

$$J_{0Z} = Z^2/2 + [3Z/(64\beta) - 5Z/8]/(1 + \gamma^2), \quad \beta = 0.845, \quad \gamma = 1/5. \tag{5.4}$$

Table 5.1 presents a comparison of the binding energies, calculated by Eq. (5.4) with the corresponding experimental values $J_{\exp}$. It is seen that there is an excellent agreement for He, but some discrepancy for two-electron ions. The discrepancy is due to the fact that $J_{0Z}$ did not take into account the effect of plasma electrons, which is important for charged two-electron systems. In accordance

Table 5.1 Comparison of ionization potentials of the ground state of the He atom and He-like ions.

| Atom ion | He $Z = 2$ | Li(+1) $Z = 3$ | Be(+2) $Z = 4$ | B(+3) $Z = 5$ | C(+4) $Z = 6$ | N(+5) $Z = 7$ | O(+6) $Z = 8$ | F(+7) $Z = 9$ | Ne(+8) $Z = 10$ |
|---|---|---|---|---|---|---|---|---|---|
| $J_{0Z}$, eV | 24.6 | 77.7 | 158.1 | 265.6 | 400.4 | 562.4 | 751.2 | 968.0 | 1211.6 |
| $J_{1Z}$, eV | | 74.5 | 153.7 | 260.2 | 393.9 | 554.8 | 742.9 | 958.2 | 1200.7 |
| $J_{\exp}$, eV [5.15] | 24.6 | 75.6 | 153.9 | 259.4 | 392.0 | 552.1 | 739.3 | 953.9 | 1195.8 |

to the Boltzmann law, the electron density near each He-like ion of the nuclear charge $Z$ increases to

$$N_e(r) = N_{e0} \exp[(Z - 2)e^2/(rT_e)]. \tag{5.5}$$

Due to this polarization effect, the ionization energy of the multi-charged ion decreases because of the exchange collisions of the 1s electrons with the plasma electrons. With the allowance for this effect, the binding energy is calculated to be

$$J_{1Z} = Z^2/2 + [3Z/(256\beta) - 5Z/8]/(1 + \gamma^2), \quad \beta = 0.845, \quad \gamma = 1/5. \tag{5.6}$$

From Table 5.1 it is seen that the binding energy, calculated by Eq. (5.6) is in a good agreement with the corresponding experimental values $J_{\exp}$ for a wide range of charged two-electron systems.

Chapter 6

# The Last Observed Line in the Spectral Series of Hydrogen Lines in Magnetized Plasmas: Revision of Inglis–Teller Concept

An isolated hydrogen atom emits practically infinite series of spectral lines, corresponding to radiative transitions from the upper level of the principal quantum number $n$ to the lower level of the principal quantum number $n_0$. (Here and below by "hydrogen atoms" and "hydrogen spectral lines" we mean atoms and spectral lines of hydrogen, deuterium, and tritium.) In other words, there is practically no restriction on how high the number $n$ can be.

However, when a hydrogen atom is placed in a uniform electric field $\mathbf{E}$, there occurs the Stark splitting of the energy levels. The splitting scales as $\sim n^2$. Therefore, at sufficiently large $n$, denoted as $n_{\max}$, the splitted adjacent levels of the principal quantum numbers $n$ and $n + 1$ merge into a quasicontinuum and the spectral series terminates. This happens when the sum of halfwidths of the Stark multiplets of the two adjacent lines becomes equal to the unperturbed separation of these two lines.

At the electric field $E$, for the multiplet of the principal quantum number $n \gg 1$, the separation $\Delta\omega(n)$ of the most shifted Stark sublevel from the unperturbed frequency $\omega_0(n)$ is $\Delta\omega(n) = 3n^2\hbar E/(2m_e e)$. Then the sum of the "halfwidths" of the two adjacent Stark multiplets of the principal quantum numbers $n$ and $n + 1$ is

$$\Delta\omega_{1/2}(n) = 3n^2\hbar E/(m_e e). \qquad (6.1)$$

The unperturbed separation (in the frequency scale) between the hydrogen spectral lines, originating from the highly-excited levels $n$ and $n + 1$ is

$$\omega_0(n + 1) - \omega_0(n) = m_e e^4/(n^3 \hbar^3). \tag{6.2}$$

By equating Eqs. (6.1) and (6.2) one finds

$$n_{\text{max}}^5 E = E_{\text{at}}/3 = 1.714 \times 10^7 \text{ CGS}, \quad E_{\text{at}} = m^2 e^5/\hbar^4 \tag{6.3}$$

($E_{\text{at}} = 1.714 \times 10^7$ CGS $= 5.142 \times 10^9$ V/cm is the atomic unit of electric field).

Inglis and Teller [6.1] applied this concept to hydrogen atoms in plasmas. In plasmas there are typically two kinds of electric fields of significantly different statistical properties: the electron microfield and the ion microfield. For broad ranges of plasma parameters and of the $n$-values, hydrogen atoms perceive the ion microfield as quasistatic. The ion microfield $E_i$ has a distribution over the ensemble of radiators. For relatively low electron densities $N_e$ (equal to the ion density $N_i$ in hydrogen plasmas) and/relatively high temperatures, the distribution is the one derived by Holtsmark [6.2].

Inglis and Teller [6.1] used the most probable field of the Holtsmark distribution, which they estimated as $E_{i\,\text{max.}} = 3.7 e N_i^{2/3} = 3.7 e N_e^{2/3}$, and obtained from Eq. (6.3) the following relation:

$$N_e n_{\text{max}}^{15/2} = 0.027/a_0^3 = 1.8 \times 10^{23} \text{ cm}^{-3}, \tag{6.4}$$

where $a_0$ is the Bohr radius. We note that Hey [6.3], by using a more accurate value of the most probable Holtsmark field $E_{i\,\text{max}} = 4.18 e N_e^{2/3}$, obtained a slightly more accurate numerical constant in the right side of Eq. (6.4), namely $0.0225/a_0^3$, while Griem [6.4] suggested this constant to be even twice smaller.

Inglis–Teller relation (6.4) constituted a simple method for measuring the electron density by the number $n_{\text{max}}$ of the observed lines of a hydrogen spectral series. The simplicity of this method is the reason why, despite the existence of more sophisticated (but more demanding experimentally) spectroscopic methods for measuring $N_e$, this method is still used in both laboratory and astrophysical plasmas. For example, Welch *et al.* [6.5] used it (with the constant in the right side of Eq. (6.4) suggested by Griem [6.4]) for determining the electron density in the low-density discharge at the tokamak Alcator C-Mod. (Tokamaks are a type of plasma machines designed for the research in the area of magnetically-controlled thermonuclear fusion leading to a practically-inexhaustible source of energy.)

However, in magnetized plasmas radiating hydrogen atoms moving with the velocity **v** across the magnetic field **B** experience a Lorentz electric field $\mathbf{E}_L = \mathbf{v} \times \mathbf{B}/c$ in addition to other electric fields. The Lorentz field $E_L$ can significantly exceed the most probable Holtsmark field $E_{\max}$. Indeed, the average Lorentz field is

$$\langle E_L \rangle = B v_T / c = 4.28 \times 10^{-3} B [T(K)]^{1/2}, \qquad (6.5)$$

where $v_T = (2T/M)^{1/2}$ is the atomic thermal velocity. It can exceed the most probable ion microfield $E_{\max}$ when the magnetic field $B$ exceeds the following critical value:

$$B_c = 4.69 \times 10^{-7} N_e^{2/3} / [T(K)]^{1/2}, \qquad (6.6)$$

(in Eqs. (6.5) and (6.6), $B$ is in Tesla and the atomic temperature $T$ is in Kelvin). For edge plasmas of tokamaks, such as, e.g., a low-density discharge in Alcator C-Mod [6.5], where $N_e \sim 3 \times 10^{13} \, \text{cm}^{-3}$ and $T \sim 5 \times 10^4$ K, we get $B_c = 4$ T, while the actual magnetic field was 8 T.

Another example can be the solar chromosphere, where the typical plasma parameters are $N_e \sim 10^{11} \, \text{cm}^{-3}$ and $T \sim 10^4$ K (except solar flares where $N_e$ can be higher by two orders of magnitude) — see, e.g., [6.6, 6.7]. In this case from Eq. (10) we get $B_c = 0.2$ T. A more accurate estimate for this example can be obtained by taking into account that non-thermal velocities $v_{\text{non-th}}$ in the solar chromosphere can be $\sim$ several tens of km/s, so that the total velocity $v_{\text{tot}} = (v_T^2 + v_{\text{non-th}}^2)^{1/2} \sim (15 - 30)$ km/s. Then $E_L = E_{i\,\max}$ already at $B \sim 0.05$ T, while B can reach 0.4 T in sunspots.

In this kind of situations, the number $n_{\max}$ of the last observable hydrogen line will not be controlled by the electron density, but rather by different parameters, as shown below. Let us first conduct a simplified reasoning along the approach of Inglis and Teller [6.1].

By substituting $E = E_{\mathrm{L}}$ in the left side of Eq. (6.3), we obtain the following relation

$$n_{\max}^{10} B^2 T(K) = 1.78 \times 10^{18} M/M_p \quad \text{or}$$
$$n_{\max}^{10} B^2 T(\mathrm{eV}) = 1.54 \times 10^{14} M/M_p, \tag{6.7}$$

where $B$ is the magnetic field in Tesla; $M$ and $M_p$ are the atomic and proton masses, respectively.

These results obtained by us in paper [6.8] constitute the breakdown of the Inglis–Teller concept for magnetized plasmas and represent a new diagnostic method allowing to measure the product $T^{1/2}B$. In other words, this method allows measuring the atomic temperature $T$, if the magnetic field is known, or the magnetic field $B$, if the temperature is known.

More accurate relations [6.8] can be derived after calculating Lorentz-broadened profiles of high-$n$ Balmer lines. Since the velocity $\mathbf{v}$ has a distribution, so does the Lorentz field $\mathbf{E}_L = \mathbf{v} \times \mathbf{B}/c$, thus resulting in the broadening of spectral lines.

The Lorentz field $\mathbf{E}_L$ is confined in the plane perpendicular to $\mathbf{B}$ where it has the following distribution

$$W_{\mathrm{L}}(E_{\mathrm{L}}) dE_{\mathrm{L}} = (2E_{\mathrm{L}}/E_{\mathrm{LT}}^2) \exp(-E_{\mathrm{L}}^2/E_{\mathrm{LT}}^2) dE_{\mathrm{L}}, \quad E_{\mathrm{LT}} = v_{\mathrm{T}} B/c. \tag{6.8}$$

Here $E_{\mathrm{LT}}$ is the average Lorentz field expressed via the thermal velocity $v_{\mathrm{T}}$ of the radiating atoms. The distribution $W_{\mathrm{L}}$ actually reproduces the shape of the two-dimensional Maxwell distribution of atomic velocities in the plane perpendicular to $\mathbf{B}$.

The Lorentz-broadened profile of a Stark component of a hydrogen line reproduces the shape of the Lorentz field distribution from Eq. (6.8)

$$S_{\alpha\beta}(\Delta\omega) = (2\Delta\omega/\Delta\omega_{L\alpha\beta}^2) \exp(-\Delta\omega^2/\Delta\omega_{L\alpha\beta}^2),$$
$$\Delta\omega_{\mathrm{L}\alpha\beta} = kX_{\alpha\beta} Bv_{\mathrm{T}}/c. \tag{6.9}$$

Here

$$k = 3\hbar/(2m_e e), \quad X_{\alpha\beta} = n_\alpha(n_1 - n_2)_\alpha - n_\beta(n_1 - n_2)_\beta, \tag{6.10}$$

where $n_1$, $n_2$ are the parabolic quantum numbers, and $n$ is the principal quantum numbers of the upper (subscript $\alpha$) and lower (subscript $\beta$) Stark sublevels involved in the radiative transition.

To ensure that for each particular hydrogen line, the profile will be *universal*, i.e., applicable for any $B$ and $T$, we chose the argument $\gamma$ of the profiles as

$$\gamma = c\Delta\omega/(kBv_T). \tag{6.11}$$

We calculated the corresponding profiles as follows:

$$P(\gamma) = P_\pi(\gamma) + P_\sigma(\gamma), \tag{6.12}$$

$$P_\pi(\gamma) = \Sigma_{(\alpha,\beta)\pi}(J_{\alpha\beta}/|X_{\alpha\beta}|)f_\pi W(\gamma/|X_{\alpha\beta}|), \tag{6.13}$$

$$P_\sigma(\gamma) = \Sigma_{(\alpha,\beta)\pi}(J_{\alpha\beta}/|X_{\alpha\beta}|)f_\sigma W(\gamma/|X_{\alpha\beta}|), \tag{6.14}$$

where

$$W(u) = 2u\exp(-u^2) \tag{6.15}$$

(Eq. (6.15) is the renormalized Eq. (6.9)). Here $J_{\alpha\beta}$ is the relative intensity of the corresponding Stark component. In Eq. (6.13) the summation is performed over $\pi$-components (i.e., components polarized along the Lorentz field), in Eq. (6.14) — over $\sigma$-components (i.e., components polarized perpendicular to the Lorentz field).[1] The quantities $f_\pi$, $f_\sigma$ in Eqs. (6.13) and (6.14) are

$$f_\pi = 1, \quad f_\sigma = 1/2, \tag{6.16}$$

for the observation parallel to **B**, or

$$f_\pi = 1/2, \quad f_\sigma = 3/4, \tag{6.17}$$

for the observation perpendicular to **B**.

As an example, Fig. 6.1 presents calculated universal Lorentz-broadened profiles of the Balmer line $H_{18}$ for the observations perpendicular to **B** (solid line) and parallel to **B** (dashed line).

---

[1]For hydrogen lines having the component of $X_{\alpha\beta} = 0$ (which is usually the $\sigma$-component), the sum in (6.19) should be complemented by $J_0\delta(\Delta\omega)$, where $J_0$ is the relative intensity of the component. Physically this means that the contribution of this component to the overall profile is controlled by other broadening mechanisms (typically, by the Doppler broadening). However, first, for 50% of hydrogen lines, such as those with even values of $(n_\alpha - n_\beta)$, there is no component of $X_{\alpha\beta} = 0$. Second, for high-$n$ hydrogen lines $(n_\alpha \gg n_\beta)$ with odd values of $(n_\alpha - n_\beta)$, the contribution of this component to the overall profile is relatively small and can be neglected.

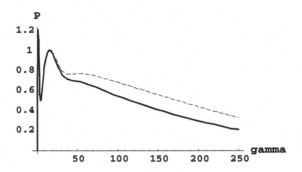

Fig. 6.1. Calculated universal Lorentz-broadened profiles $P(\gamma)$ of the Balmer line $H_{18}$ for the observations perpendicular to **B** (solid line) and parallel to **B** (dashed line). The argument $\gamma = c\Delta\omega/(kBv_T)$, where $k = 3\hbar/(2m_e e)$. The intensities of both profiles are relative to the intensity of the 2nd maximum (whose intensity is set as unity). The 1st maximum would be smeared out by the Doppler broadening.

From the calculated profiles we derived the following practically important result for highly excited Balmer lines. For any two adjacent high-$n$ Balmer lines (such as, e.g., $H_{16}$ and $H_{17}$, or $H_{17}$ and $H_{18}$), the sum of their half widths at half maximum in the frequency scale, the sum being denoted here simply as $\Delta\omega_{1/2}$ , turned out to be

$$\Delta\omega_{1/2} = A[3n^2\hbar Bv_T/(2m_e ec)], \qquad (6.18)$$

where the constant $A$ depends on the direction of observation as follows:

$$A = 0.80 \text{ (observation perpendicular to } \mathbf{B}), \qquad (6.19)$$

$$A = 1.00 \text{ (observation parallel to } \mathbf{B}), \qquad (6.20)$$

$$A = 0.86 \text{ (``isotropic'' observation).} \qquad (6.21)$$

Here by the "isotropic" observation we describe the situation where along the line of sight there are regions with various directions of the magnetic field, which could be sometimes the case in astrophysics.

Thus, the above formulas, by using the observable quantity $n_{\max}$, allow to measure the atomic temperature $T$, if the magnetic field is known, or the magnetic field $B$, if the temperature is known. The accuracy can be increased by taking into account additional broadening mechanisms as follows:

$$[(\Delta\omega_{1/2}(n))^2 + \Delta\omega_Z^2 + (\Delta\omega_D)_{1/2}^2]^{1/2} = m_e e^4/(n^3\hbar^3), \qquad (6.22)$$

where the Zeeman width $\Delta\omega_Z$ and the Doppler width $(\Delta\omega_D)_{1/2}$ are

$$\Delta\omega_Z = eB/(2m_e c), \quad (\Delta\omega_D)_{1/2} = 2(\ln 2)^{1/2}\omega_0 v_T/c = 1.665\omega_0 v_T/c. \tag{6.23}$$

As we noted in [6.8], the Lorentz and Zeeman mechanisms can be combined together "exactly" using the fact that the problem of a hydrogen atom in the crossed electric and magnetic fields allows an analytical solution which is exactly within the subspace spanned on the states of the principal quantum number $n$. This fact, being the consequence of the $O(4)$ symmetry of the hydrogen atom (and of the corresponding Kepler problem), was presented already in 1927 by Born in frames of the "old quantum theory" [6.9] and was later elaborated in more detail in the contemporary quantum theory by Demkov, Monozon, and Ostrovsky [6.10]. The characteristic frequencies, arising in this solution, are $\xi\Omega$, where

$$\Omega = \{[\Delta\omega_L/(n-1)]^2 + \Delta\omega_Z^2\}^{1/2}, \quad \xi = -(n-1), -(n-2), \ldots, (n-1), \tag{6.24}$$

$\Delta\omega_L$ and $\Delta\omega_Z$ being given in Eq. (6.23). It is seen that the characteristic frequencies $\xi\Omega$ are similar to the left side of Eq. (6.22) if in the latter the Doppler broadening would be disregarded.

For example, for the low-density discharge at Alcator C-Mod [6.5], where deuterium Balmer lines up to $n_{\max} = 16$ were observed and the magnetic field was $B = 8\,$T, we obtain from Eq. (6.22): $T = 6\,$eV.

Another example: in solar chromosphere, where Balmer lines up to $n_{\max} \sim 30$ have been observed, assuming $T \sim 10^4$K and the average non-thermal velocity $v_{\text{non-th}} \sim 20\,$km/s, we obtain from Eq. (6.22): $B \sim 0.2\,$T.

# Extrema in Transition Energies Resulting Not in Satellites But in Dips Within Spectral Lines

## 7.1. Breaking the Paradigm and Revealing Charge-Exchange-Caused Dips (x-dips)

It is well-known that the shape of spectral lines of a radiating atom/ion (radiator) in a gas or in a plasma depends on the energy terms of the combined quantum system "radiator + perturber(s)". Frequently the energy difference between the terms involved in the radiative transition, being plotted versus the radiator–perturber separation, shows extrema. This situation has been studied for over 30 years — both theoretically and experimentally (see, e.g., [7.1–7.11] and references therein). The paradigm based on these studies is that the extrema in the transition energy result in *satellites* in spectral line profiles [7.1–7.11].

Here we present the results first published in paper [7.12] showing that the extrema in the transition energy can also result in *dips* in spectral line profiles. Moreover, we demonstrate that for the practically important case where the extremum in the transition energy is due to the *charge exchange*, its spectral signature *most probably should be a dip* rather than a satellite.

We consider a radiative transition between two terms corresponding to some Stark component. We use atomic units and therefore employ the same notation $f(R)$ for both the transition energy and the transition frequency; $R$ is the distance between the radiator and the perturbing atom or ion. We denote as $g(R)$ the area-normalized probability distribution of the quantity $R$. In the quasistatic approximation, the area-normalized profile $I(\Delta\omega)$ of the Stark component versus the detuning $\Delta\omega$ from the unperturbed frequency $\omega_0$ is usually given by

$$I(\Delta\omega) = \int_0^\infty dR G(R)\delta[\Delta\omega - f(R)], \quad G(R) = g(R)J(R)/J(\infty),$$
(7.1)

where $J(R)$ is a frequency-integrated relative intensity of the Stark component.

We consider a vicinity of some particular distance $R_0$ corresponding to a small part of the component profile around $\Delta\omega_0 = f(R_0)$. In a relatively simple case where $f(R)$ does not have an extremum at $R = R_0$, from Eq. (7.1) one usually obtains

$$I(\Delta\omega_0) = G(R_0)/|f'(R_0)|.$$
(7.2)

However, if $f(R)$ has an extremum at $R = R_0$, so that the first derivative $f'(R)$ vanishes, then Eq. (7.2) as well as Eq. (7.1) at $\Delta\omega = \Delta\omega_0$ becomes inapplicable. Physically this means that some feature in the profile may arise in the vicinity of $\Delta\omega_0$. For obtaining a finite value of $I(\Delta\omega_0)$, one should allow for additional broadening mechanisms and substitute the delta-function in Eq. (7.1) by a more realistic profile, such as a Lorentzian or a Gaussian. For example, for the Lorentzian having a full-width at half-maximum (FWHM)$\gamma$, as shown in Sec. 3.3.2 of the book [7.13], the intensity $I(\Delta\omega_0)$ becomes:

$$I(\Delta\omega_0) \approx G(R_0) \int_{-\infty}^\infty d(R - R_0)[(\gamma/(2\pi)]/\{(\gamma/2)^2$$
$$+ [f''(R_0)/2]^2(R - R_0)^4\}$$
$$= G(R_0)\{2/[\gamma|f''(R_0)|]\}^{1/2}.$$
(7.3)

The result (7.3) was obtained under the assumption that the function $G(R)$ is the slowest out of two factors in the integrand. The validity of this assumption will be analyzed later on.

We are interested primarily in the situation where an extremum in the energy difference between an upper term $a$ and a lower term $a_0$

is caused by an avoided crossing of the radiator's term a with some perturber's term $a'$. We denote as $f_-(R)$ the transition frequency between the *original* terms $a$ and $a_0$ at $R \leq R_0$ and as $f_+(R)$ the transition frequency between the *original* terms $a'$ and $a_0$ at $R \geq R_0$. Here by *"original"* we mean the terms as they would be if the terms $a$ and $a'$ would not be coupled and therefore would cross (Fig. 7.1).

In the vicinity of $R = R_0$, as a result of the avoided crossing, there occurs a transition of the energy difference from $f_-(R)$ to $f_+(R_0)$ as well as a transition of the slope from $f'_-(R_0)$ to $f'_+(R_0)$. If $f'_-(R_0)$ and $f'_+(R_0)$ have the opposite signs, the avoided crossing causes an extremum in the energy/frequency difference (Fig. 7.1).

Combining Eqs. (7.2) and (7.3), it is easy to calculate the ratio of the true intensity $I(\Delta\omega_0)$ to the original intensity $I_0(\Delta\omega_0)$ (that is the intensity $I_0(\Delta\omega_0)$ which would be if it were no coupling of the $a$ and $a'$ terms):

$$I(\Delta\omega_0)/I_0(\Delta\omega_0) = |f'_-(R_0)|\{2/[\gamma|f''(R_0)|]\}^{1/2}. \qquad (7.4)$$

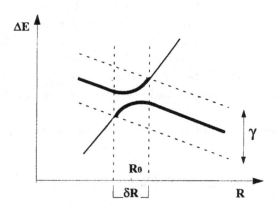

Fig. 7.1. Transition energies $f_-(R) = E_a - E_{a0}$ and $f_+(R) = E_{a'} - E_{a0}$ versus the radiator–perturber separation R, plotted in a vicinity $\delta R$ of an avoided crossing of the perturber's term $a'$ with the radiator's term $a$ at $R = R_0$ in the course of the radiative transition from the term $a$ to the term $a_0$. The transition energy $f_-(R)$ actually occupies a band of a width $\gamma$ (shown by dashed lines) controlled primarily by the dynamical broadening caused by electron and ion microfields in a plasma. The radiator's transition energy modified by the avoided crossing is shown by the bold line. In the interval $\delta R$, the transition energy has two branches, corresponding to the fact that the wave function of the radiator's term in this interval is a linear combination of wave functions of two different energies.

Here we come to the following *central point*. From Eq. (7.4) it is clear that if $|f'_-(R_0)|$ would be relatively large and/or $\gamma|f''(R_0)|$ would be relatively small, we would have $I(\Delta\omega_0)/I_0(\Delta\omega_0) > 1$, thus indicating the formation of a peak (satellite). However, if $|f'_-(R_0)|$ is relatively small and/or $(\gamma|f''(R_0)|)$ is relatively large, we have $I(\Delta\omega_0)/I_0(\Delta\omega_0) < 1$, thus indicating the formation of a *dip*, rather than a satellite. This result *disproves the existing paradigm* in accordance to which any extremum in the transition energy/frequency could manifest only as a satellite.

Physically, the earlier findings can be illustrated as follows. Let us consider again the starting formula (7.1) for $I(\Delta\omega)$, but with the delta-function substituted by a more realistic profile (e.g., by the Voigt profile). If the extremum in the transition energy/frequency is "gradual" (i.e., $|f''(R_0)|$ is relatively small), then in the course of integration over $R$ around $R_0$, the frequency remains in the vicinity of $\Delta\omega_0$ "longer" than at the absence of the extremum. Therefore, more intensity gets accumulated at $\Delta\omega_0$ than it would be at the absence of the extremum, thus resulting in a peak (satellite). However, if the extremum in the transition energy/frequency is "sharp" (i.e., $|f''(R_0)|$ is relatively large), then in the course of integration over $R$ around $R_0$, the frequency remains in the vicinity of $\Delta\omega_0$ "shorter" than at the absence of the extremum. Therefore, less intensity gets accumulated at $\Delta\omega_0$ than it would be at the absence of the extremum, thus resulting in a dip.

The absolute value of the second derivative $f''(R_0)$ can be estimated as

$$f''(R_0) \approx [f'_+(R_0) - f'_-(R_0)]/\delta R, \tag{7.5}$$

where $\delta R$ is the interval where there occurs the conversion of the original radiator's term $a$ into $a'$. The interval $\delta R$ can be found from the following considerations.

Due to the dynamical Stark broadening and the radiative broadening (the latter resulting in a "natural" width), the radiator has a finite lifetime $1/\gamma$. Consequently, the transition energy of the radiator actually *occupies a band of the width* $\gamma$ (see Fig. 7.1). Therefore, it is easy to find that

$$\delta R = \gamma / |f'_+(R_0) - f'_-(R_0)|. \tag{7.6}$$

Thus the ratio $I(\Delta\omega_0)/I_0(\Delta\omega_0)$ can be represented in the form:

$$I(\Delta\omega_0)/I_0(\Delta\omega_0) \approx 2^{1/2} |f'_-(R_0)/[f'_+(R_0) - f'_-(R_0)]|. \tag{7.7}$$

Clearly, the right hand side of Eq. (7.7) is controlled by the inverse value of the *relative change of the derivative of the transition energy* at the crossing and does not depend on the dynamical Stark width $\gamma$. If the relative change of this derivative is small, there forms a satellite. However, if the relative change of this derivative is large, we find again that the *existing paradigm breaks down*: there forms a *dip* (rather than a satellite).

From now on we focus here specifically at the avoided crossings in the system of two Coulomb centers (dicenters) $ZeZ'$. We would like to remind that terms of the dicenter may indeed cross: the well-known non-crossing rule [7.14] is inapplicable because the system possesses an algebraic symmetry higher than the geometrical symmetry [7.15]. (The corresponding additional conserved quantity is the projection of the generalized Runge–Lenz vector [7.16] on the internuclear axis.) However, due to the charge exchange some of the crossings transform into avoided crossings [7.17, 7.18].

We consider a radiative transition in a hydrogen/hydrogen-like atom/ion of the nuclear charge $Z$ at the presence of the nearest perturber which is a fully-stripped ion of the charge $Z' \neq Z$ located at the distance $R$. The upper (term $a$) and the lower (term $a_0$) states involved in the radiative transition have the principal quantum numbers $n$ and $n_0$, respectively. At some distance $R_0$, a $Z$-term of the principal quantum number $n$ (term $a$) experiences an avoided crossing with a $Z'$-term of the principal quantum number $n'$ (term $a'$).

We are going to show that when an extremum in the transition energy is due to a charge-exchange-caused avoided crossing, which occurs at a relatively large distance

$$R \gg \max(n^2/Z, n'^2/Z'), \tag{7.8}$$

then *practically always it results in a dip* in the profile of the corresponding Stark component of the spectral line, rather than in a satellite. In other words, we are going to show that in this situation, practically always we have $I(\Delta\omega_0)/I_0(\Delta\omega_0) < 1$.

Indeed, the condition (7.8) allows to treat separately a set of $Z$-terms (perturbed by the $Z'$-ion) and a set of $Z'$-terms (perturbed by the $Z$-ion) as well as to use $1/R$-expansion for both the energies $(E_a, E_{a'}, E_{a0})$ of the corresponding terms $(a, a', a_0)$ and the transition energies $f_-(R)$ and $f_+(R)$. The expansion of energies up to the $1/R^2$ — terms has the form

$$E_a = -Z^2/(2n^2) - Z'/R + 3Z'nq/(2ZR^2),$$
$$E_{a'} = -Z'^2/(2n'^2) - Z/R + 3Zn'q'/(2Z'R^2), \qquad (7.9)$$
$$E_{a0} = -Z^2/(2n_0^2) - Z'/R + 3Z'n_0q_0/(2ZR^2),$$

where $q$, $q'$, $q_0$ are the electric quantum numbers of the terms $a$, $a'$, $a_0$, respectively (e.g., $q = n_1 - n_2$, where $n_1$, $n_2$ are the parabolic quantum numbers, and so on). Then for the transition energies we obtain the following expressions:

$$f_-(R) = E_a - E_{a0} = (Z^2/2)(n_0^{-2} - n^{-2}) + 3Z'(nq - n_0q_0)/(2ZR^2),$$
$$f_+(R) = E_{a'} - E_{a0} = [Z^2/(2n_0^2) - Z'^2/(2n'^2)] + (Z' - Z)/R$$
$$+ 3(n'q'Z/Z' - n_0q_0Z'/Z)/(2R^2). \qquad (7.10)$$

Finally, for the derivatives of the transition energies, we find the following first nonvanishing terms:

$$f'_-(R_0) = -3Z'(nq - n_0q_0)/(ZR^3),$$
$$f'_+(R) = -(Z' - Z)/R^2 - 3(n'q'Z/Z' - n_0q_0Z'/Z)/R^3. \qquad (7.11)$$

It is seen from Eq. (7.11) that, given the condition (7.8), the absolute value of the slope of $f_+(R)$ is much greater than the absolute value of the slope of $f_-(R)$. Further, we substitute the expressions (7.11) in Eq. (7.7) and obtain:

$$I(\Delta\omega_0)/I_0(\Delta\omega_0) \approx [2^{1/2}3Z'/(ZR_0)]|(nq - n_0q_0)/(Z - Z')|. \qquad (7.12)$$

Thus, in view of the condition (7.8), we have indeed proven the earlier statement that $I(\Delta\omega_0)/I_0(\Delta\omega_0) \ll 1$, which means the formation of a dip. We call this charge-exchange caused dip an "x-dip".

The result (7.12) is applicable to the Stark components of $nq - n_0 q_0 \neq 0$ (i.e., for lateral Stark components, which constitute the majority of the components). For the Stark components of $nq - n_0 q_0 = 0$ (central components), the energies $E_a$ and $E_{a0}$ in Eq. (7.9) should be complemented by the quadrupole term ($\sim 1/R^3$). Thereafter, we obtain for the central components:

$$I_c(\Delta\omega_0)/I_{0c}(\Delta\omega_0) \approx 3Z'(n^4 - n^2 - n_0^4 - n_0^2)/(2^{1/2}Z^2|Z - Z'|R_0^2). \tag{7.13}$$

Obviously, in view of the condition (7.8), we have

$$I_c(\Delta\omega_0)/I_{0c}(\Delta\omega_0) \ll I(\Delta\omega_0)/I_0(\Delta\omega_0) \ll 1, \tag{7.14}$$

which means the formation of the x-dip. Moreover, Eq. (14) shows that x-dips in the profile of the central components should be even more pronounced than in the profiles of the lateral components.

As a measure of the "contrast" of the x-dips, we introduce a quantity called "visibility" defined as follows:

$$V \equiv \left| 1 - \left[ I_k J_k(\infty) + \sum_{i \neq k} I_{0,i} J_i(\infty) \right] \middle/ \sum_{\text{all } i} I_{0,i} J_i(\infty) \right|$$

$$= \left| (1 - I_k/I_{0k}) \middle/ \left[ \sum_{\text{all } i} I_{0,i} J_i(\infty)/I_{0k} J_k(\infty) \right] \right|$$

$$\approx \left| (1 - I_k/I_{0k}) \middle/ \left[ \sum_{\text{all } i} J_i(\infty)/J_k(\infty) \right] \right|. \tag{7.15}$$

Here $I_k J_k(\infty)$ is the intensity of the Stark component # $k$ at $\Delta\omega = \Delta\omega_0$, in whose profile the x-dip occurs; $I_{0,i} J_i(\infty)$ is the original intensity of the Stark component # $i$ of this spectral line at $\Delta\omega = \Delta\omega_0$.

We note that for fast estimates of the visibility of x-dips in hydrogen like lines consisting of a large number of components, one can use the following result deduced from our analysis of the tables of intensities of Stark components [7.19]: The total number of the *strong* Stark components of a hydrogenlike spectral line $n \leftrightarrow n_0$ is approximately equal to $n_0(2n - 1)$. Therefore, the ratio of the frequency-integrated intensities in Eq. (7.15) can be estimated as

$$\sum_{\text{all } i}[J_i(\infty)/J_k(\infty)] \approx n_0(2n - 1). \tag{7.16}$$

There are upper and lower limits which determine the range of electron densities, where the x-features can be observed. The upper limit $N_e^{\text{upper}}$ physically comes from the condition that the dynamical broadening should not be so large as to wash out the x-feature. Mathematically, for the derivation of $I(\Delta\omega_0)$ in Sec. 7.1.2 to be valid, it is required that the characteristic interval $\Delta R_G$ of a significant change of the "slow" function $G(R)$ would be greater than the characteristic interval $\Delta R_\gamma$ of a significant change of the "fast" quasi-Lorentzian. The latter quantity is

$$\Delta R_\gamma = (\gamma/|f''(R_0)|)^{1/2} = \gamma/|f'_+(R_0) - f'_-(R_0)|, \tag{7.17}$$

while the interval $\Delta R_G$ can be defined through the averaged absolute value of the quantity $G'(R)/G(R)$:

$$\Delta R_G = \left[\int_0^\infty dR G(R)|G'(R)/G(R)|\right]^{-1}$$

$$= \left[\int_0^\infty dR|G'(R)|\right]^{-1} = [2G(R_{\text{peak}})]^{-1}, \tag{7.18}$$

where $R_{\text{peak}}$ is the location of the peak of $G(R)$. The last equality in Eq. (7.18) holds in the usual case where $G(R)$ has only one peak. Thus, the requirement $\Delta R_\gamma < \Delta R_G$ translates into the condition:

$$\gamma < |f'_+(R_0) - f'_-(R_0)|/[2G(R_{\text{peak}})]. \tag{7.19}$$

We emphasize that for the x-feature (i.e., x-dip or x-satellite) to be clearly visible, the change of the derivative of the transition energy at the crossing $|f'_+(R_0) - f'_-(R_0)|$ should belong to a certain range. Indeed, by combining Eqs. (7.7) and (7.19), we obtain the following.

For the x-dip to be clearly visible, the restriction on $|f'_+(R_0) - f'_-(R_0)|$ is

$$|f'_+(R_0) - f'_-(R_0)| \gg \max[2^{1/2}|f'_-(R_0)|, \quad 2\gamma G(R_{\text{peak}})]. \tag{7.20}$$

For the x-satellite to be clearly visible, the corresponding range is

$$2^{1/2}|f'_-(R_0)| \gg |f'_+(R_0) - f'_-(R_0)| \gg 2\gamma G(R_{\text{peak}}). \qquad (7.21)$$

The lower limit of electron densities $N_e^{\text{lower}}$ physically comes from the requirement that the crossing distance $R_0$ should not differ too much from the most probable interperturber distance (otherwise, $R_0$ would belong to a very low-weight part of the function $G(R)$. More rigorously, the ratio of the original intensity of the Stark component at $\Delta\omega_0$ to its peak intensity should exceed certain threshold

$$I_0(\Delta\omega_0)/\max_{\Delta\omega}[I_0(\Delta\omega)] > \varepsilon, \qquad (7.22)$$

what can be re-written, using Eqs. (7.1) and (7.2), as

$$G(R_0)|f'_-(R_0)|^{-1}/[G(R_{\text{peak}})|f'_-(R_{\text{peak}})|^{-1}] > \varepsilon, \qquad (7.23)$$

where, say $\varepsilon \sim 0.1$.

For two lines with very strong central components, that is, for $L_\alpha$ and $H_\alpha$, while finding the lower limit for $N_e$, the following additional requirement should be also checked

$$(\gamma/\Delta\omega_0)^2 > \varepsilon, \qquad (7.24)$$

where $\varepsilon \sim 0.1$ and $\Delta\omega_0$ is the location of the x-feature. This is again to ensure that there is a sufficient intensity at the location of the x-feature. The reason that the condition (7.24) is required in addition to the condition (7.23) is that the full width at the half maximum of these two lines is usually controlled by the dynamical broadening due to electrons and ions, rather than by the broadening due to quasistatic ions incorporated in the condition (7.23). Thus, for these two lines, the lower limit for $N_e$ should be determined as

$$N_e = \max[N_e^{(7.23)}, N_e^{(7.24)}], \qquad (7.25)$$

where $N_e^{(7.23)}$ is the lower limit deduced from Eq. (7.23) and $N_e^{(7.24)}$ is the lower limit deduced from Eq. (7.24).

In experimental studies of x-dips at various plasmas (some of which are presented in Sec. 7.4), in each particular experiment x-dips were observed in the range of the electron densities of no more than two orders of magnitude — consistent with the analytical results of this section.

For applying our theory to particular $ZeZ'$-systems it is important to use the following selection rule for the charge exchange in such systems. For each $Z'$-term $(n'_1, n'_2, m')$, the charge exchange, also manifested as an avoided crossing, is possible with no more than one $Z$-term, namely: either with the Z-term of the following parabolic quantum numbers [7.17, 7.18]

$$n_1 = n'_1, \quad m = m', n_2 = n - n'_1 - |m'| - 1 \qquad (7.26)$$

or not at all, if the set $(n_1, n_2, m)$ given by Eq. (7.26) does not correspond to any $Z$-term. Physically, this selection rule follows from the picture of the charge exchange as the corresponding interaction of states in two adjacent potential wells (one — centered at the charge $Z$, another — at the charge $Z'$) and from the fact that for such interaction to be possible, the radial wave functions of these states should have the same number of nodes [7.17, 7.18].

## 7.2. Classical Model of x-Dips

In Chapter 3 we presented a classical analytical solution for a quasimolecule consisting of two stationary Coulomb centers of charges $Z$ and $Z'$, separated by a distance $R$, and one electron. We showed that for the circular orbits of the electron, there are three energy terms of the same symmetry, i.e., of the same projection of the angular momentum on the internuclear axis. Further, we found that two of these classical energy terms undergo a $V$-type crossing (the crossing resembling the inclined letter $V$) — see Fig. 3.3.

A classical intensity of the radiation of the electron executing a circular orbit is given by $I = 2\rho F^2/(3c^3)$ [7.20], where $F$ is the angular frequency, which in our case is:

$$F(w, b) = (Z^{1/2}/R^{3/2})f(w, b) = (Z^2/M^3)m_0^3(w, b)f(w, b),$$
$$(7.27)$$

where $f(w, b)$ and $m_0(w, b)$ are given by Eqs. (3.30) and (3.13), respectively. Both the radiation frequency $F$ and the radiation intensity $I$ depend on the internuclear distance $R$ and vary as $R$ varies. So, in a plasma medium, the averaging over the ensemble of the distances $R$ yields in the following classical spectral line profile

$$S(\Omega) = \int_0^\infty dR h(R, R_0) I(R) \delta\{[\Omega - F(R)] M^3/Z^2\}. \qquad (7.28)$$

Here $h(R, R_0)$ is the distribution function of the distances $R$; $R_0 = [3/(4\pi N)]^{1/3}$ is the mean interperturber distance, $N$ is the perturber's density. We introduce a scaled (dimensionless) radiation frequency $\omega \equiv (M^3/Z^2)\Omega$, a scaled (dimensionless) mean interperturber distance $r_0 \equiv (Z/M^2)R_0$ and use the binary distribution function (see, e.g., [7.21]) $h = (3R^2/R_0^3)\exp[-(R/R_0)^3] = (3Z/M^2)r_0^{-3} m_0^{-4}(w, b)\exp[-r_0^{-3} m_0^{-6}(w, b)]$. As a result, for any pair of $b$ and $r_0$, we obtain the dependence of the classical spectral line profile $S$ on $\omega$ in the following parametric form (via $w$):

$$S(w, b, r_0) = [4Z/(c^3 M)]s(w, b, r_0), \quad \omega = m_0^3(w, b)f(w, b),$$
$$s(w, b, r_0) \equiv \sum r_0^{-3} m_0^{-9/2}(w, b) f^{3/2}(w, b) |m_0'(w, b)/$$
$$[3m_0\prime(w, b)f(w, b) + m_0(w, b)f\prime(w, b)]|$$
$$\times \exp[-r_0^{-3} m_0^{-6}(w, b)], \qquad (7.29)$$

where the sign ⁄ (prime) stands for the derivative with respect to the scaled coordinate $w$. The parametrically defined dependence of $s$ on $\omega$ is, generally, double-valued, and the sign $\Sigma$ in Eq. (7.29) means the summation over both branches of $s$. The dependence of $S$ on $\omega$ is a classical spectral profile of the central ("unshifted") Stark components, corresponding to the radiative transitions between the Stark states of q = 0 of the hydrogen like of the nuclear charge $Z$.

Figure 7.2 shows the dependence of the scaled (dimensionless) spectral intensity s on the scaled radiation frequency $\omega$ for $b = 3$, $r_0 = 5.5$. The most remarkable feature is a minimum (dip) at $\omega \approx 0.74$. The dip corresponds to the value of $w$ in a vicinity of $w_c$, where the classical energy terms cross. This means that the physical mechanism of the formation of the dip is an enhancement of charge exchange

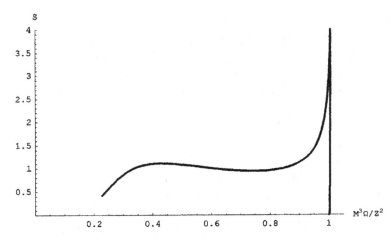

Fig. 7.2. The dependence of the scaled classical spectral intensity of the radiation $s = [c^3 M/(4Z)]S$ on the scaled radiation frequency $\omega = (M^3/Z^2)\Omega$ for $Z' = 3Z$ and for the scaled mean interperturber distance $r_0 = 5.5$.

in the vicinity of the crossing, as described in detail in Sec. 3.3. In other words, *we reproduced classically the spectral signature of charge exchange.* We note that in Fig. 7.2, as the scaled radiation frequency $\omega$ diminishes, the scaled spectral intensity $s$ is terminated at some value $\omega_{\min}$ corresponding to the point in the upper classical energy term, where the scaled increment of the instability $|\omega_-|$ becomes equal to the scaled angular frequency $f$ of the (primary) $\varphi$-motion.

We emphasize that the above example represents a typical situation. Actually, *for any pair of $Z$ and $Z' \neq Z$, some feature (a dip or at least a shoulder) always occurs* in the classical spectral line of the ion having the smallest out of $Z$ and $Z'$. (For $Z' = Z$ there is no feature in the classical spectral line — as should be expected.) We also note that quantum mechanics does not allow to analytically reproduce the spectral profile around the dip, while classical mechanics and electrodynamics does it easily. Particularly, the classical approach allowed us to incorporate a variation of the transition dipole moment with $R$ in an exact analytical fashion, which would be impossible in the quantum approach. However, a direct comparison of these classical results with experiments is limited to radiative transitions between highly-excited states.

## 7.3. Advanced Quantal Theories of x-Dips

In the previous sections of this chapter, the following two sets of questions have not been addressed.

1. What is the shape of the x-dip? Is it a structure consisting of the dip itself surrounded by two adjacent bumps — as it was in the case of the Langmuir-wave-caused dips (abbreviated as L-dips)? We note that theoretical and experimental studies of the L-dips, summarized in book [7.13], showed that their positions significantly depend on the electron density $N_e$ of plasmas, while the positions of the x-dips are practically independent of $N_e$.
2. The mechanism of the formation of the x-dip suggested in paper [7.22] was a sharp decrease of the lifetime $t_{\text{life}}$ of the upper state of the radiating ion (hereafter, the radiator) at some internuclear distance $R = R_{\text{cr}}$, while in the previous sections of this chapter the mechanism was described as a sharp change of the slope of the transition energy at $R = R_{\text{cr}}$ — as the incident ion switches from one quasimolecular term to another at their avoided crossing (also known as, anticrossing). What is the relation between these two explanations? Do they describe the same processes or they represent two independent mechanisms working simultaneously?

Here we show that the x-dip should be indeed a structure consisting of the dip itself surrounded by two adjacent bumps. We also show that the explanation provided in paper [7.22] and the explanation given in the previous sections of this chapter are two independent mechanisms working simultaneously. Based on the obtained results, we provide a *method for determining the rates of charge exchange* from the experimental shape of the x-dips.

We use atomic units and therefore employ the same notation $f(R)$ for both the transition energy and the transition frequency of the radiative transition. We denote as $g(R)$ the area-normalized probability distribution of the internuclear $R$.

Far from the anticrossing, the lifetime $t_{\text{life}}$ of the upper state of the radiator is controlled by the inelastic part of the dynamical broadening by the electron and ion microfields, as well as by the radiative broadening: $t_{\text{life}}(R) = 1/\gamma_{\text{nonCE}}(R)$, where $\gamma_{\text{nonCE}}(R) = \gamma_{\text{Stark}}(R) + \gamma_{\text{rad}}(R)$. Here $\gamma_{\text{Stark}}(R)$ is the frequency of inelastic collisions with electrons and ions leading to virtual transitions from the upper state of the radiator to other states (usually, the dominant contribution to $\gamma_{\text{Stark}}(R)$ is due to electronic collisions), $\gamma_{\text{rad}}(R)$ is the radiative width. The radiative width (which depends on $R$ via the $R$-dependence of the dipole matrix elements) is expected to be relatively small — so, we would still call the contribution $\gamma_{\text{nonCE}}(R)$ collisional. The quantity $\gamma_{\text{nonCE}}(R)$ varies very slowly away from $R_{\text{cr}}$.

In the vicinity of the anticrossing, there exists an additional channel for the decay of the upper state of the radiator: charge exchange. A sharp decrease of the lifetime due to charge exchange is controlled by the frequency of charge-exchange-causing collisions $\gamma_{CE}(R) = N_i \langle v \sigma_{\text{CE}}(v) \rangle$, where $N_i$ is the density of incident ions, $\langle v \sigma_{\text{CE}}(v) \rangle$ is the rate coefficient of charge exchange.

With the allowance for charge exchange, the lifetime of the upper state of the radiator becomes $t_{\text{life}}(R) = 1/\gamma_t(R)$. Here $\gamma_t(R)$ is the total frequency of inelastic collisions equal to $\gamma_{CE}(R) + \gamma_{\text{nonCE}}(R)$.

Since the lifetime of the upper state of the radiator is at the focus of one of the two mechanisms of the x-dip formation, an appropriate starting formula for the lineshape should be such chosen as to contain explicitly the lifetime of the upper state of the radiator $t_{\text{life}}(R)$ or the total frequency of inelastic collisions $\gamma_t(R)$. It is well-known that the term $\exp[-\gamma_t(R)\tau]$ in the correlation function corresponds to a Lorentzian of the FWHM equal to $\gamma_t(R)$. Therefore, for the area-normalized profile $I(\Delta\omega)$ of the Stark component versus the detuning $\Delta\omega$ from the unperturbed frequency $\omega_0$ we use the following expression

$$I(\Delta\omega) = \int_0^\infty dR\, G(R) L[\Delta\omega - f(R)], \quad G(R) = g(R) J(R)/J(\infty),$$
$$(7.30)$$

where $J(R)$ is a relative intensity of the Stark component and $L(x)$ is the Lorentzian[1]:

$$L(x) = [\gamma_t/(2\pi)]/[(\gamma_t/2)^2 + x^2]. \qquad (7.31)$$

---

[1]The separation of (quasi) static and dynamic ions is done as follows [7.23]. For a given value $\tau$ of the argument of the correlation function $C(\tau)$, the ion-dynamical part of $C(\tau)$

We consider a vicinity of some particular distance $R_0$ corresponding to a small part of the component profile around $\Delta\omega_0 = f(R_0)$.

In the previous sections of this chapter, for deriving the intensity drop at the center of the x-dip, in the integral in Eq. (7.1) we used the Taylor expansion of the transition energy $f(R)$ at the point $R_0 = R_{cr}$ corresponding to a possible extremum of $f(R)$ at the anticrossing. Therefore, the term (in the Taylor expansion) proportional to the first derivative $f''(R_0)$ was zero. As a result we obtained in paper [7.3]:

$$I(\Delta\omega_0) = G_0\{2/[\gamma_t|f_0''|]\}^{1/2}, \quad G_0 \equiv G(R_0), \quad f_0'' \equiv f''(R_0). \quad (7.32)$$

Now we rewrite this result by introducing instead of $\Delta\omega_0$ its scaled (dimensionless) counterpart $\Omega_0$:

$$\Omega_0 \equiv \Omega(R_0) = 2\Delta\omega_0/\gamma_t = 2f(R_0)/\gamma_t. \quad (7.33)$$

Then the corresponding scaled (dimensionless) intensity $I_{s0}(\Omega_0)$ at the center of the x-dip can be represented as

$$I_{s0}(\Omega_0) \equiv \gamma_t I(\Delta\omega_0)/2 = G_0\{\gamma_t/[2|f_0''|]\}^{1/2}. \quad (7.34)$$

In distinction to the above, now in the integral in Eq. (7.1) we expand $f(R)$ at some point $R_0$ located in the vicinity of the extremum of $f(R)$, but *not necessarily coinciding with the point of the extremum.* Therefore, the term proportional to the first derivative $f''(R_0)$ will be now different from zero.

We introduce a scaled (dimensionless) counterpart b of this first derivative:

$$b \equiv 2f_0'/(\gamma_t|f_0''|)^{1/2}, \quad f_0'' \equiv f'(R_0). \quad (7.35)$$

After some elementary transformation, we now obtain the following formula for the scaled intensity $I_s(\Omega_0)$,

$$I_s(\Omega_0) = I_{s0}(\Omega_0)j_{00}(b), \quad (7.36)$$

---

originates from the collisions, for which the instants of the closest approach $t_0$ fall within the interval $(-\tau/2, \tau/2)$. The rest of the perturbing ions are considered static. This is consistent with the fact that at $\tau \to \infty$ all ions are dynamic, while at $\tau \to 0$ all ions are static. Thus, to different values of $\tau$ correspond different proportions of dynamic and static ions. Among static ions, the nearest neighbor (to the radiator) together with the radiator is considered as a quasimolecule (the dynamical effect involving the nearest neighbor ion is limited to charge exchange and is reflected by the quantity $\gamma_{CE}(R)$). Dynamic ions contribute to the frequency of inelastic collisions leading to virtual transitions from the upper state of the radiator to other states.

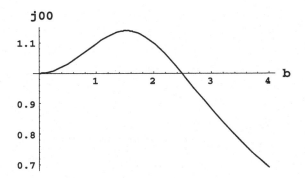

Fig. 7.3. Dependence of the quantity $j_{00}$ defined by Eq. (7.37) on the dimensionless parameter $b$ defined by Eq. (7.35).

where $j_{00}(b)$ is a universal function defined as

$$
j_{00}(b) \equiv (2^{1/2}/\pi) \int_{-\infty}^{\infty} dy (1 + 2y^2 + y^4)/
$$
$$
[1 + (2 + b^2)y^2 + 2by^3 + 2(1 + b^2)y^4 + 4by^5
$$
$$
+ (2 + b^2)y^6 + 2by^7 + y^8]. \tag{7.37}
$$

This integral can be integrated analytically. However, the resulting expression is bulky and we omit it.

Figure 7.3 shows the plot of $j_{00}$ versus $b$. It is seen that as $R_0$ moves away from $R_{\mathrm{cr}}$ (i.e., away from the anticrossing), so that $|f_0'|$ and b increase, the intensity goes through a maximum — both at the red and blue sides of the dip. Thus we predict *two bumps surrounding the x-dip*.

Now we study the second set of questions outlined in Sec. 7.3.1. In the previous sections of this chapter, the total frequency of inelastic collisions $\gamma_t$ was considered as a constant (calculated at the center of the dip). Let us relax this assumption.

The two terms within $\gamma_t(R) = \gamma_{\mathrm{CE}}(R) + \gamma_{\mathrm{nonCE}}(R)$ significantly differ by their dependence on the internuclear distance $R$: $\gamma_{\mathrm{CE}}(R)$ rapidly decreases away from the anticrossing $R_{\mathrm{cr}}$, while $\gamma_{\mathrm{nonCE}}(R)$ varies very slowly away from $R_{\mathrm{cr}}$. We introduce the following notations:

$$
\gamma \equiv \gamma_{\mathrm{nonCE}}(R_{\mathrm{cr}}), \quad \gamma_{\mathrm{CEo}} \equiv \gamma_{\mathrm{CE}}(R_0), \quad a \equiv \gamma_{\mathrm{CEo}}/\gamma. \tag{7.38}
$$

First, we set $R_0 = R_{cr}$, so that $b = 2f_0'/(\gamma|f_0''|)^{1/2} = 0$ (because $f_0' = 0$ at $R_0 = R_{cr}$), and we can study the second set of questions independent of the first one. By approximating the shape of $\gamma_{CE}(R)$ in the vicinity of $R_{cr}$ as a Lorentzian, after some elementary transformation, we obtain the following formula for the scaled intensity $I_s(\Omega_0)$,

$$I_s(\Omega_0) = I_{s0}(\Omega_0)j_0(a), \tag{7.39}$$

where $j_0(a)$ is a universal function defined as

$$j_0(a) \equiv (2^{1/2}/\pi) \int_{-\infty}^{\infty} dy[1 + a + (2 + a)y^2 + y^4]/$$
$$[(1 + a)^2 + (2 + 2a)y^2 + 2y^4 + 2y^6 + y^8]. \tag{7.40}$$

The integral (7.40) can also be integrated analytically, like the integral in Eq. (7.37). However, the resulting expression is also bulky and we omit it.

Figure 7.4 shows the plot of $j_0$ versus $a$. It shows that as the ratio $a = \gamma_{CEo}/\gamma$ increases, the x-dip becomes deeper — compared to its depth at $a = 0$.

This means that there are actually *two independent mechanisms of the formation of the x-dip.* One of them is a sharp increase of the total frequency of inelastic collisions $\gamma_t$ at $R = R_{cr}$, another — a sharp change of the slope of the transition energy at $R = R_{cr}$. *They*

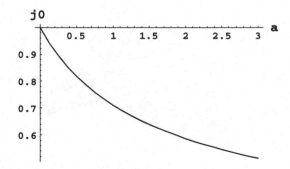

Fig. 7.4. Dependence of the quantity $j_0$ defined by Eq. (7.40) on the dimensionless parameter a defined by Eq. (7.38).

*are not two descriptions of the same*: they can work independently and *enhance each other*.

Now we combine the results of the Secs. 7.3.2 and 7.3.3 and consider a general case, where both $a = \gamma_{CEo}/\gamma \neq 0$ and $b = 2f_0'/(\gamma|f_0''|)^{1/2} \neq 0$ (we remind that $f_0' \neq 0$ means that $R_0 \neq R_{cr}$). It turns out that in the general case, the scaled intensity $I_s(\Omega_0)$ can be obtained in the form

$$I_s(\Omega_0) = I_{s0}(\Omega_0)j(a,b). \tag{7.41}$$

Here $j(a,b)$ is the following universal function of two variables

$$j(a,b) \equiv (2^{1/2}/\pi) \int_{-\infty}^{\infty} dy f(y,a,b), \tag{7.42}$$

where

$$\begin{aligned}
f(y,a,b) \equiv [1 + a + (2+a)y^2 + y^4]/ \\
[(1+a)^2 + (2+2a+b^2)y^2 + 2by^3 + 2(1+b^2)y^4 \\
+ 4by^5 + (2+b^2)y^6 + 2by^7 + y^8].
\end{aligned} \tag{7.43}$$

The universal function $j(a,b)$ reduces to $j_{00}(b)$ for $a = 0$ and to $j_0(a)$ for $b = 0$. It is the even function of $b$: $j(a,-b) = j(a,b)$.

Figure 7.5 shows a plot of $j(a,b)$ versus $b$ as $a$ varies from 0 to 5 with a step of 0.25. From Fig. 7.5 it is seen that as the parameter $a$

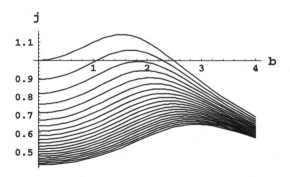

Fig. 7.5. Plot of the function j defined by Eqs. (7.42), (7.43) versus the dimensionless parameter b (defined by Eq. (7.35)) while the dimensionless parameter $a$ (defined by Eq. (7.38)) varies from 0 (the top curve) to 5 (the bottom curve) with a step of 0.25.

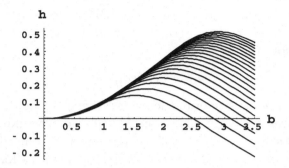

Fig. 7.6. Plot of the function $h$ defined by Eq. (7.44) versus the dimensionless parameter $b$ (defined by Eq. (7.35)) while the dimensionless parameter a (defined by Eq. (7.38)) varies from 0 (the bottom curve) to 5 (the top curve) with a step of 0.25.

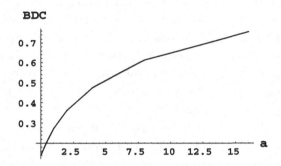

Fig. 7.7. BDC versus the dimensionless parameter a (defined by Eq. (7.38)). The BDC is defined as the maximal value of the function $h(a, b)$ for any given value of $a$ (see Fig. 7.6).

increases, the bumps move further away from the center of the x-dip and their intensity decreases.

For studying a *relative* variation of $j(a, b)$, it is convenient to introduce another function:

$$h(a, b) \equiv j(a, b)/j(a, 0) - 1. \qquad (7.44)$$

For any given $a$, it shows a relative change of the intensity away from the center of the x-dip. Figure 7.6 shows a plot of $h(a, b)$ versus $b$ as a varies from 0 to 5 with a step of 0.25.

For any given $a = \gamma_{\text{CEo}}/\gamma$, we will call the maximal value of $h(a, b)$ as the Bump-to-Dip Contrast (BDC). From Fig. 7.6 it is seen that the BDC is function of $a$. Figure 7.7 shows the BDC versus $a$.

Thus, using the dependence of the BDC versus $a = \gamma_{\text{CEo}}/\gamma$, *we can determine the ratio $\gamma_{CE}(R_0)/\gamma$ from the experimentally measured BDC.* We note that

$$\gamma_{\text{CEo}} = N_i \langle v\sigma_{\text{CE}}(v) \rangle, \tag{7.45}$$

where $N_i$ is the density of incident ions, $\langle v\sigma_{CE}(v) \rangle$ is the rate coefficient of charge exchange. Consequently, after determining the experimental value $a_{\text{exp}} = \gamma_{\text{CEo}}/\gamma$ from the experimental BDC, *we can then deduce the rate coefficient of charge exchange* as follows:

$$\langle v\sigma_{\text{CE}}(v) \rangle = \gamma a_{\text{exp}}/N_i. \tag{7.46}$$

The quantity $\gamma \equiv \gamma_{\text{nonCE}}(R_{\text{cr}})$ in Eq. (7.46), representing the frequency of inelastic collisions with electrons and ions leading to virtual transitions from the upper state of the radiator to other states, can be calculated for given plasma parameters $N_e$, $N_i$, $T_e$, and $T_i$ by using one of few contemporary theories (presented, e.g., in book [7.23]).

We note that the formalism chosen here for describing the shape of x-dips seems to be the most convenient. However, it is conceivable that some alternative formalism(s) might be also developed for the same purpose. For example, the dynamical phenomenon of the disappearance (with some probability) of the upper state of the radiator in the vicinity of the anticrossing might be attempted to be described not through the charge-exchanged-caused contribution $\gamma_{CE}$ to the collisional width, but by "de-freezing" the motion of the nearest neighbor ion and introducing a significant reduction of the dipole matrix elements $\mathbf{d}_{\alpha\beta}$ between the upper state $\alpha$ and the lower state $\beta$. However, the latter would become dependent on *all* parameters of ionic collisions, including the velocity of the incident (nearest neighbor) ion. This might lead to some conceptual (and computational) problems in calculating the corresponding lineshape.

Indeed, the basic formula for the lineshape contains the operator $Q(\tau) = \langle \mathbf{dd}U(\tau/2, -\tau/2)\rangle_v$, where $U$ is the evolution operator in the interaction representation, $\langle \ldots \rangle_v$ stands for the velocity average. Far from the anticrossing, the dipole moment operator $\mathbf{d}$ does not depend on the velocity: the operator $Q$ simplifies to $Q(\tau) = \mathbf{dd} < U(\tau/2, -\tau/2) >_v$ and the product $\mathbf{dd}$ translates, in particular, in intensities $J_{\alpha\beta}$ of various Stark components. However, in the vicinity of the anticrossing, due to the velocity dependence of the dipole moment operator $d$, the product $dd$ cannot be factored out of the velocity averaging in $Q(\tau) = \langle ddU(\tau/2, -\tau/2)\rangle_v$. Therefore, even if one would calculate the quantities $\langle dd\rangle_v$ and still call them "intensities of Stark components", these quantities would not be directly related to the lineshape calculations.

The results presented in the previous sections of this chapter were obtained analytically — in the parabolic coordinates using the multi-pole expansion for the interaction of the hydrogenic ion of the nuclear charge $Z$ (or $Z'$) with the fully-stripped ion of the charge $Z'$ (or $Z$), and without taking into account the plasma electron screening. We retained a large number of terms in the multi-pole expansion to achieve the accuracy adequate for the comparison with experiments.

In paper [7.24] we calculated the positions of the x-dips via numerical simulations — by diagonalizing the Hamiltonian corresponding to the exact interaction in the elliptical coordinates. Then we incorporated the electron screening model into the simulations. It turned out that the usage of the exact ionic interaction shifted the x-dips positions slightly further to the red, while the allowance for the electron screening shifted them in the opposite direction. As a result, the theoretical positions of the x-dips still had zero or little dependence on the plasma electron density $N_e$.

## 7.4. Practical Applications: Experimental Determinations of the Rates of Charge Exchange Between Multicharged Ions Using Observed x-Dips

The first observation of the x-dips was made in Germany at the gas-liner pinch [7.22]: the x-dip was found in the blue side of the $H\alpha$ line emitted by hydrogen atoms $H(Z = 1)$ perturbed by fully stripped helium $He(Z' = 2)$ (Fig. 7.8). The dip was observed for a relatively small range of electron densities around $10^{18}$ cm$^{-3}$. For lower densities the quasi-crossing distance $R_c$ was much lower than the

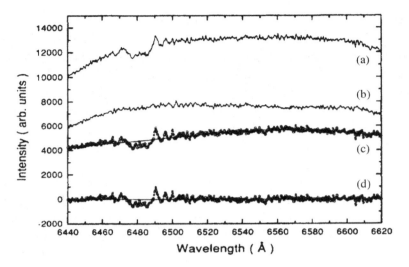

Fig. 7.8. Blue side of the $H_\alpha$ line of hydrogen with (a) and without (b) injecting hydrogen about 140 ns before the maximum compression of the gas-liner pinch [7.22]. The electron density is $N_e = 1.79 \times 10^{18}$ cm$^{-3}$, the temperature $T_e = 6.7$ eV. Spectrum (c) shows the difference between (a) and (b); in (d) additionally a Lorentzian profile as the best fit [solid line in (c)] is subtracted.

mean inter-ionic distance $R_i$ and for higher densities the dynamical broadening $\gamma_{\mathrm{nonCE}}(R_c)$ was smoothing the dip structure, both reasons being unfavorable for experimental observations. In these experiments the electron density $N_e$ was measured by Thomson scattering and it was verified that the positions of the X-dips did not depend on $N_e$.

Later the x-dips were observed in laser-produced plasmas characterized by a high electron density $10^{22} - 3 \times 10^{22}$ cm$^{-3}$ [7.25]. The setup was implemented at the laser facility LULI in France using a nanosecond laser at $10^{14}$W/cm$^2$ and a high-resolution Vertical-geometry Johann Spectrometer ($R = 8000$). The targets used for the observation of the x-dips were aluminum carbide Al$_4$C$_3$ strips inserted in carbon substrate. The emission from the heterogeneous plasma made up of Al and C ions exhibited spectroscopic signatures of charge exchange. X-dips were observed for the first time in the experimental profile of the Ly$\gamma$ line of Al XIII ($Z = 13$) perturbed by fully stripped carbon CVI ($Z' = 6$), as shown in Fig. 7.9. The

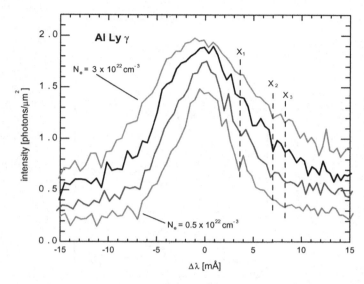

Fig. 7.9. Observation of the X-dip structures $X_1$, $X_2$, $X_3$ in the red wing of the Al XIII Ly$\gamma$ line perturbed by fully-stripped carbon in laser-produced plasma experiments at LULI [7.25].

positions of the dips did not vary significantly in the small density domain. For smaller densities the spectral line profile was too narrow to allow the visibility of the x-dip structures and at higher densities the dips were smoothed out, as explained in Sec. 7.1.5.

From the experimental bump-to-dip ratio, the rate of charge exchange between a hydrogenic aluminum ion in the state $n = 4$ and a fully stripped carbon was found to be: $\langle v\sigma_{CE}(v)\rangle = (5.2 \pm 1.1)10^{-6}$ cm$^3$/s. This is an important fundamental reference date virtually inaccessible by other experimental methods.

The next experimental study of x-dips was done in Prague: in plasma-wall interaction experiments performed at PALS laser facility [7.26]. A plasma jet of aluminum ions, produced by the nanosecond iodine laser of the intensity $3 \times 10^{14}$ Wcm$^{-2}$ incident on a foil, interacted with a massive carbon target. The high spectral and spatial resolution Vertical-geometry Johann Spectrometer was adjusted for the observation of X-dips in the experimental profile of the Ly$\gamma$ line of Al XIII ($Z = 13$) (Fig. 7.10). Two x-dips clearly visible in the red

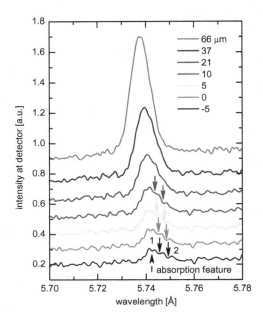

Fig. 7.10. Observation of the x-dips in the red wing of the Al XIII Ly$\gamma$ line in the plasma-wall interaction experiments at the PALS laser facility [7.26]. A plasma jet of aluminum ions interacted with a massive carbon target. The spectra correspond to different distances from the target surface (and thus to different electron densities): from $-5$ $\mu$m (the bottom curve) to $+66$ $\mu$m (the top curve).

wing are the signatures of charge exchange accompanying the plasma-wall interaction. Simulations of the electron density showed that it could reach $5 \times 10^{22}$ cm$^{-3}$, which is higher than the electron density achieved at LULI at the study of the same line (Fig. 7.9). A weak dependence of the positions of the x-dips with the density has been detected — consistent with the simulation from paper [7.24]. The dip, visible in the far-red wing, is a Langmuir-wave-caused dip [7.13] corresponding to a possible multi-quantum resonance, thus providing a spectroscopic diagnostic of the electron density.

The latest experimental study of x-dips was done in Japan using *femtosecond* laser-driven *cluster*-based plasma [7.27]. The experiments were performed at Kansai Photon Science Institute. Two Ti:sapphire laser facilities were employed: wavelength approximately 800 nm, pulse duration 40 fs. The clusters were created when a gas of

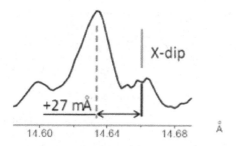

Fig. 7.11. Spectrum of the Ly$\varepsilon$ line of O VIII perturbed by He$^{+2}$, obtained in femtosecond laser-driven cluster-based experiments [7.27]. The laser intensity was $4 \times 10^{17}$ W/cm$^2$. The experimental x-dip is marked by the solid vertical line.

a high initial pressure was expanded into vacuum through a specially designed supersonic nozzle, which consisted of three coaxial conical surfaces. $CO_2$-clusters with a diameter of about 0.5 mm for pure $CO_2$ (gas pressure 20 bar) and 0.22 mm for the mixed gas of 90% He + 10% $CO_2$ (gas pressure 60 bar) were produced in both experiments. The spectral resolution of a focusing spectrometer was ~4000.

From the theoretical point of view, we found a *unique opportunity*. For the line Ly$\varepsilon$ of O VIII perturbed by He$^{+2}$, the three possible x-dips should occur practically at the same location in the profile — merging into one *"x-superdip"* located at $(30 \pm 3.5)$ mA in the red wing. This x-dip is clearly visible in shot #2: it is marked by the solid vertical line in Fig. 7.11. The central minimum of the experimental x-dip is at 27 mA (in a good agreement with the theory) and is surrounded by two bumps, as predicted by the theory.

From the experimental BDC $a_{\exp}$ of the x-dip, we obtained the rate of charge exchange between the ion O$^{+8}$ in the state of $n = 6$ and the ion He$^{+2}$ in the state of $n' = 2$: $\langle v\sigma_{CE}(v) \rangle = (1.5 \pm 0.3) \times 10^{-6}$ cm$^3$/s. This is yet another new fundamental data virtually inaccessible by other experimental methods.

## 7.5. Future Prospects

Up to now the x-dips were observed only in 3 spectral lines (H$\alpha$ line of hydrogen perturbed be He$^{++}$, Ly$\gamma$ line of Al XIII perturbed by C$^{6+}$, and Ly$\varepsilon$ line of O VIII perturbed by He$^{++}$). In paper [7.28]

Table 7.1 Prospective hydrogen-like lines and solid targets for observing x-dips in laser-produced plasmas [7.28].

| Line | Perturber | Reciprocal case | Target |
|------|-----------|-----------------|--------|
| N VII L-gamma 19.83 A | $Li^{+3}$ | Li III L-alpha 135.0 A perturbed by $N^{+7}$ | Lithium nitride $Li_3N$ (solid) |
| Al XIII L-beta 6.053 A | $Be^{+4}$ | | Beryllium aluminum alloy |
| Na XI L-gamma 8.021 A | $B^{+5}$ | B V L-alpha 48.59 A perturbed by $Na^{+11}$ | Borax $Na_2B_4(H_2O)_{10}$ (mineral/crystals) |
| S XVI L-beta 3.991 A | $B^{+5}$ | | Boron sulfide $B_2S_3$ (polymeric material) |
| Al XIII L-gamma 5.739 A | $C^{+6}$ | C VI L-alpha 33.74 A perturbed by $Al^{+13}$ | Aluminium carbide $Al_4C_3$ (crystals) or Aluminium carbonate $Al_2(CO_3)_3$ (component of mineral strontiodesserite) |
| K XIX L-beta 2.826 A | $C^{+6}$ | | Potassium carbonate $K_2CO_3$ (salt) |
| P XV L-gamma 4.31 A | $N^{+7}$ | N VII L-alpha 24.78 A perturbed by $P^{+15}$ | Phosphorus nitride $P_3N_5$ (amorphous solid) |
| Ti XXII L-beta 2.105 A | $N^{+7}$ | | Titanium nitride TiN (ceramic material) |
| Mn XXV L-beta 1.627 A | $O^{+8}$ | | Manganese dioxide $MnO_2$ (solid) |
| K XIX L-gamma 2.680 A | $F^{+9}$ | F IX L-alpha 14.99 A perturbed by $K^{+19}$ | Potassium bifluoride $KHF_2$ (salt) or Potassium fluoride KF (mineral carobbite) |
| Ni XXVIII L-beta 1.293 A | $F^{+9}$ | | Nickel fluoride $NiF_2$ (coating that forms on nickel alloys) |

we presented 16 prospective hydrogen-like spectral lines and 11 corresponding solid targets for future studies of the x-dips, shown as follows.

Even more important is the following result from paper [7.28]: the prediction that x-dips should be possible to observe not only in spectral lines of hydrogen-like ions, but also in spectral lines of He-like ions. In the previous studies the x-dip phenomenon was considered to be possible only in spectral lines of hydrogen atoms and hydrogen-like ions: due to the existence of *exact* algebraic symmetries relevant only to hydrogenic systems and to the corresponding two-Coulomb-center systems having one electron. Exact algebraic (i.e., higher than geometrical) symmetries of these quantum systems lead to exact additional conserved quantities, known as Runge–Lenz vector and its generalizations (see. e.g., the latest paper [7.16] on this subject).

The concept of symmetry is very general: in fact, any approximate analytical theory can be considered as a simplified version of a more complicated problem, the simplification being achieved by using some *approximate* symmetry [7.29]. Actually, any regularity in the energy spectrum of a quantum system reflects certain symmetry properties [7.29].

While helium-like ions and the corresponding two-Coulomb-center systems having two electrons do not possess *exact* additional conserved quantities, they possess *approximate* additional conserved quantities [7.30]. This should be sufficient for the x-dips to occur also in these systems.

In more detail, the higher than geometrical symmetry of two-Coulomb-center systems having one electron is manifested by an additional conserved quantity (integral of motion) — in addition to the energy $E$ and the angular momentum projection $M$. This additional conserved quantity is the projection (on the internuclear axis) of the super-generalized Runge–Lenz vector [7.16]

$$\mathbf{A} = \mathbf{p} \times \mathbf{L} - L^2/R\mathbf{e_z} - Z\mathbf{r}/r - Z'(\mathbf{R} - \mathbf{r})/|\mathbf{R} - \mathbf{r}| + Z'\mathbf{e_z},$$
$$\mathbf{e_z} = \mathbf{R}/R, \tag{7.47}$$

where $\mathbf{p}$ and $\mathbf{L}$ are the linear and angular momenta vectors, respectively; $\mathbf{r}$ is the radius vector of the electron. In the two-electron case, after substituting in Eq. (7.47) $Z$ by $Z_{\text{eff}} = Z - 1$ and treating $\mathbf{r}$ as the radius vector of the outer electron, the projection of the vector $\mathbf{A}$ on the internuclear axis is an approximately conserved quantity.

This analytical theory (as well the corresponding analytical theory for the one-electron case presented in Secs. 7.1 and 7.3) is valid for a typical situation where the internuclear distance $R_c$, at which a quasicrossing occurs, satisfies the following conditions. The first condition is

$$R_c \gg \max(n^2/Z_{\text{eff}}, n'^2/Z'), \qquad (7.48)$$

i.e., for $R_c$ to significantly exceed the characteristic sizes of the two corresponding hydrogenic subsystems. Under this condition, one can use the multipole expansion for the energy terms. The second condition is

$$R_c > \max[(3n^5 Z'/Z_{\text{eff}}^3)^{1/2}, \quad (3n'^5 Z_{\text{eff}}/Z'^3)^{1/2}]. \qquad (7.49)$$

Table 7.2 Prospective He-like lines and reciprocal cases of H-like lines, and solid targets for observing x-dips in laser-produced plasmas.

| Line | Perturber | Reciprocal case | Target |
|------|-----------|-----------------|--------|
| O VII He-gamma 17.78 A | $Li^{+3}$ | Li III L-alpha 135.0 A perturbed by $O^{+7}$ | Lithium oxide (lithia) $Li_2O$ (solid) |
| Si XIII He-beta 5.681 A | $Be^{+4}$ | | Be-Si binary alloy or beryllium-doped silicon |
| Mg XI He-gamma 7.473 A | $B^{+5}$ | B V L-alpha 48.59 A perturbed by $Mg^{+11}$ | Boracite $Mg_3B_7O_{13}Cl$ (mineral) |
| Si XIII He-gamma 5.405 A | $C^{+6}$ | C VI L-alpha 33.74 A perturbed by $Si^{+13}$ | Silicon carbide (carborundum) SiC (solid) |
| Ca XIX He-beta 2.705 A | $C^{+6}$ | | Calcium carbonate $CaCO_3$ (the common substance found in rocks and the main component of shells of marine organisms, snails, and egg shells) |
| S XV He-gamma 4.089 A | $N^{+7}$ | N VII L-alpha 24.78 A perturbed by $S^{+15}$ | Ammonium sulfate $(NH_4)_2SO_4$ (crystals/granuls) |
| V XXII He-beta 2.027 A | $N^{+7}$ | | Vanadium nitride VN (solid) |
| Fe XXV He-beta 1.574 A | $O^{+8}$ | | Iron oxides FeO, $Fe_3O_4$, $Fe_2O_3$ (crystalline solid) |
| Ca XIX He-gamma 2. 57 A | $F^{+9}$ | F IX L-alpha 14.99 A perturbed by$Ca^{+19}$ | Calcium fluoride (fluorite) $CaF_2$ (mineral/solid) |
| Cu XXVIII He-beta 1.257 A | $F^{+9}$ | | Copper fluoride $CuF_2$ (crystalline solid) |

This condition is necessary to ensure the existence of the energy level $n_{\text{eff}}$ of the radiating $Z_{\text{eff}}$-ion and of the energy level $n'$ of the $Z'$-ion. Namely, the relation (7.49) is necessary to ensure that under the electric field of $Z'$-ion at the distance $R_c$, the level n of the radiating $Z_{\text{eff}}$-ion does not merge with the level $n + 1$, as well as to ensure that under the electric field of $Z_{\text{eff}}$-ion at the distance $R_c$, the level $n'$ of the $Z'$-ion does not merge with the level $n' + 1$. For practically all dicenters, the condition (7.49) is more restrictive than the condition (7.48).

The third condition puts an upper limit on $R_c$:

$$R_c < [3n^2 Z'/(Z_{eff}\Delta E)]^{1/2}, \tag{7.50}$$

where $\Delta E$ is the size of the unperturbed multiplet of the principal quantum number $n$. This inequality ensures that for the $Z$-ion state of the principal quantum number $n$, the Stark splitting caused by the $Z'$-ion at the distance $R_c$ significantly exceeds the unperturbed separation of the sublevels of the $n$-shell. This condition allows using parabolic coordinates.

It can be shown that for the restrictions on $R_c$ to be fulfilled, the ratio $Z_{\text{eff}}/Z'$ should slightly exceed (but not to be equal to) a small integer. This explains the choice of prospective candidates for observing x-dips in spectral line profiles of He-like ions from laser-produced plasmas, as presented in Table 7.2 containing 15 prospective He-like spectral lines (as well as 5 pairs of reciprocal spectral H-like spectral lines) and 10 corresponding solid targets.

These results very significantly extend the range of fundamental data on charge exchange between multicharged ions that can be obtained via the x-dip phenomenon, but not by any other method.

# Chapter 8

# Conclusions

We presented paradigm-breaking results in several areas of atomic and molecular physics. In some cases we used quantal formalism (such as in Chapters 2, 5 and 6), but in other cases we used classical formalism (such as in Chapters 3 and 4). Sometimes there is a misconception that with the advent of quantum mechanics, there is no need for classical calculations. So let us dispel this misconception.

There are numerous research areas, where a purely classical description based on first principles provided an adequate picture of the reality. In addition to the physical systems presented in Chapters 3 and 4, there are other such areas in the physics of the microscopic world: for example, the multiphoton ionization of hydrogen in a strong microwave field. Specifically, experiments in this area revolutionized the way we view the atomic physics of highly excited atoms (see, e.g., a review [8.1]). However, the interpretation of the experiments remained a puzzle until its stochastic, diffusional nature was uncovered in 1979 by the then-new theory of chaos [8.2]. In the intervening two decades this fundamental problem has been researched so intensively that it has become an "industry" often loosely called "quantum chaos" (see, e.g., [8.3–8.6] and references therein). Particularly, some new dynamical symmetries of the problem and its non-trivial celestial analogies have been revealed [8.7, 8.8].

Speaking of classical symmetries, we should mention a review by Wheeler [8.9]. Its title speaks for itself: "Not-so-classical mechanics: unexpected symmetries of classical motion". As noted in Chapter 4, for the overwhelming majority of the microscopic systems, the additional ("non-geometrical") integrals of the motions exist only classically, but not quantally (few exceptions, such as a hydrogen atom etc. had been mentioned above). Thus, *classical mechanics has an intrinsic, built-in advantage over quantum mechanics* in this regard.

The fact that classical formalism in many cases provides an adequate picture of the reality should not cast any doubt on the computational methods used for quantal, fully-numerical calculations of the same phenomena. Rather, first of all, it demonstrates the *principle of plurality*: physical phenomena in many cases can be adequately described by several models, some of them being classical. Second, a classical description usually provides a better physical insight than the quantal one. An example is the explanation of the classical non-radiating states (the analog of the quantal stationary states) via the non-Einsteinian time dilation (Chapter 4). Third, our results could revitalize classical calculations at a more sophisticated level, which would rival computational methods used in quantal numerical calculations. Fourth, they could also further stimulate the corresponding quantal calculations and their comparison with the classical results.

One of the illustrations of the latter is the phenomenon of charge-exchange-caused dips (x-dips) in profiles of spectral lines in plasmas where the nuclear charge of the radiating atom/ion differs from the charge of surrounding ions (Chapter 7). This phenomenon was discovered experimentally and explained in frames of the quantal description in paper [8.10]. Later a classical description of x-dips was provided in paper [8.11] (which gave rise to several works on one-electron Rydberg quasimolecules presented in Chapter 3). Then a whole new series of theoretical and experimental works on x-dips has been performed, specifically for laser-produced plasmas [8.12–8.18]. As a result, apart from the fundamental interest, the x-dip phenomenon became practically important as a method for the experimental determination of the rate of charge exchange between multicharged ions. This is a fundamental reference data for various applied projects, the data virtually inaccessible by other experimental methods.

We hope that paradigm-breaking results in several areas of atomic and molecular physics, presented in this book, would stimulate further advances in these research areas. Some of the future experimental and theoretical studies could be: (i) testing intimate details of the nuclear structure by performing atomic, rather than nuclear, experiments; (ii) extending classical studies on crossings of the energy terms and of charge exchange to two-electron and three-electron diatomic Rydberg quasimolecules; (iii) applying the classical generalized Hamiltonian dynamics, including the non-Einsteinian time dilation, to a broader scope of atomic and molecular systems.

# Appendix A

# Classical Description
# of Muonic-Electronic Negative
# Hydrogen Ion in Circular States

Studies of muonic atoms and molecules, where one of the electrons is substituted by the heavier lepton $\mu^-$, have several applications. The first one is muon-catalyzed fusion (see, e.g., [A.1–A.3] and references therein). When a muon replaces the electron either in the $dde$-molecule ($D_2^+$), which becomes the $dd\mu$-molecule, or in the $dte$-molecule, which becomes the $dt\mu$-molecule, the equilibrium internuclear distance becomes about 200 times smaller. At such small internuclear distances, the fusion can occur with a significant probability, which has been observed in $dd\mu$ or even with a higher rate in $dt\mu$ [A.1–A.3]. The second application is a laser-control of nuclear processes. This has been discussed in the context of the interaction of muonic molecules with superintense laser fields [A.4]. Another application is a search for strongly interacting massive particles (SIMPs) proposed as dark matter candidates and as candidates for the lightest supersymmetric particle (see, e.g., [A.5] and references therein). SIMPs could bind to the nuclei of atoms, and would manifest themselves as anomalously heavy isotopes of known elements. By greatly increasing the nuclear mass, the presence of a SIMP in the nucleus effectively eliminates the well-known reduced mass correction in a hydrogenic atom. Muonic atoms are better candidates (than electronic atoms) for observing this effect because the muon's much larger mass (compared to the electron) amplifies the reduced mass correction [A.5]. This may be detectable in astrophysical objects [A.5].

Another line of research is studies of the negative ion of hydrogen $H^-$, which can also be denoted as *epe*-system (electron–proton–electron), constitute an important line of research in atomic physics and astrophysics. It has only one bound state — the ground state having a relatively small bound energy of approximately 0.75 eV. This *epe*-system exhibits rich physics. Correlations between the two electrons are strong already in the ground state. With long-range Coulomb interactions between all three pairs of particles, the dynamics is particularly subtle in a range of energies 2–3 eV on either side of the threshold for break-up into proton + electron + electron at infinity [A.6]. There are strong correlations in energy, angle, and spin degrees of freedom, so that perturbation theory and other similar methods fail [A.6]. Experimental studies of $H^-$ provided a testing ground for the theory of correlated multielectron systems. Compared to the helium atom, the structure of $H^-$ is even more strongly influenced by interelectron repulsion because the nuclear attraction is smaller for this system [A.7]. In addition to the above fundamental importance, the rich physics of $H^-$ is also important in studies of the ionosphere's D-layer of the Earth atmosphere, the atmosphere of the Sun and other stars, and in development of particle accelerators [A.6].

Here we combine the above two lines of research: Studies of muonic atoms/molecules and studies of negative hydrogen ion. Namely, we consider whether a muonic hydrogen atom can attach an electron and become a muonic negative hydrogen ion, i.e., $\mu pe$-system. Specifically, we study a possibility of circular states in such system. We show that the muonic motion can represent a rapid subsystem, while the electronic motion can represent a slow subsystem — the result that might seem counterintuitive.

First, we find analytically classical energy terms for the rapid subsystem at the frozen slow subsystem, i.e., for the quasimolecule where the muon rotates around the axis connecting the immobile proton and the immobile electron. The meaning of classical energy terms is explained below. We demonstrate that the muonic motion is stable. We also conduct the analytical relativistic treatment of the muonic motion.

Then we unfreeze the slow subsystem and analyze a slow revolution of the axis connecting the proton and electron. We derive the condition required for the validity of the separation into the rapid and slow subsystems.

Finally we show that the spectral lines, emitted by the muon in the quasimolecule $\mu pe$, experience a red shift compared to the corresponding spectral lines that would have been emitted by the muon in a muonic hydrogen atom (in the $\mu p$-subsystem). Observing this red shift should be one of the ways to detect the formation of such muonic negative hydrogen ions.

As for physical processes leading to the formation of muonic-electronic negative hydrogen ions, one of the processes could be the following:

$$e + \mu p \to \mu pe$$

(which sometimes might be followed by the decay $\mu pe \to \mu + pe$). Such formation of the $\mu pe$-systems was discussed, e.g., in paper [A.8], where these systems were called resonances. The theoretical approach based on the separation of rapid and slow subsystems requires in this case the muon to be in a state of a high angular momentum. Luckily, the experimental methods to create muonic hydrogen atoms $\mu p$ (necessary for the above reaction) lead to the muon being in a highly-excited state (see, e.g., review [A.9] and paper [A.10]). We also mention paper [A.11] where it has been shown, in particular, that the distribution of the muon principal quantum number in muonic hydrogen atoms peaks at larger and larger values with the increase of the energy of the muon incident on electronic hydrogen atoms.

We consider a quasimolecule where a muon rotates in a circle perpendicular to and centered at the axis connecting a proton and an electron — see Fig. A.1. As we show below, in this configuration the muon may be considered the rapid subsystem while the proton and electron will be the slow subsystem, which essentially reduces the problem to the two stationary Coulomb center problem, where the effective stationary "nuclei" will be the proton and electron. The straight line connecting the proton and electron will be called here "internuclear" axis. We use the atomic units in this study.

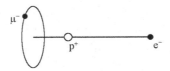

Fig. A.1. A muon rotating in a circle perpendicular to and centered at the axis connecting the proton and the electron.

Because of the difference of muon and electron masses, the muon–proton separation is much smaller than the electron–proton separation. Therefore, it should be expected that the spectral lines, emitted by this system, would be relatively close to the spectral lines emitted by muonic hydrogen atoms. In other words, the presence of the electron should result in a relatively small shift of the spectral lines (compared to muonic hydrogen atoms); however, this shift would be an important manifestation of the formation of the quasimolecule $\mu pe$.

A detailed classical analytical solution of the two stationary Coulomb center problem, where an electron revolves around nuclei of charges $Z$ and $Z'$, has been presented in Chapter 3, Sec. 2. Here we base our presentation in part on those results.

The Hamiltonian of the rotating muon is

$$H = \frac{1}{2m}\left(p_z^2 + p_\rho^2 + \frac{p_\varphi^2}{\rho^2}\right) - \frac{Z}{\sqrt{z^2 + \rho^2}} - \frac{Z'}{\sqrt{(R-z)^2 + \rho^2}}, \quad (A.1)$$

where $m$ is the mass of the muon (in atomic units $m = 206.7682746$), $Z$ and $Z'$ are the charges of the effective nuclei (in our case, $Z = 1$ and $Z' = -1$), $R$ is the distance between the effective nuclei, ($\rho$, $\varphi$, $z$) are the cylindrical coordinates, in which $Z$ is at the origin and $Z'$ is ar $z = R$, and ($p_\rho$, $p_\varphi$, $p_z$) are the corresponding momenta of the muon.

Since $\varphi$ is a cyclic coordinate, the corresponding momentum is conserved:

$$|p_\varphi| = \text{const} = L. \quad (A.2)$$

It should be noted that we denoted here the muon angular momentum as $L$, while reserving the notation $M$ for the electron angular momentum.

Substituting $|p_\varphi|$ from Eq. (A.2) into Eq. (A.1), we obtain the Hamiltonian for the $z$- and $\rho$-motions

$$H_{z\rho} = \frac{p_z^2 + p_\rho^2}{2} + U_{\text{eff}}(z, \rho), \qquad (A.3)$$

where the effective potential energy is

$$U_{\text{eff}}(z\rho) = \frac{L^2}{2m\rho^2} - \frac{Z}{\sqrt{z^2 + \rho^2}} - \frac{Z'}{\sqrt{(R-z)^2 + \rho^2}}. \qquad (A.4)$$

Because in a circular state $p_z = p_\rho = 0$, the total energy $E(z, \rho) = U_{\text{eff}}(z, \rho)$.

With $Z = 1$, $Z' = -1$ and the scaled quantities

$$w = \frac{z}{R}, \quad v = \frac{\rho}{R}, \quad \varepsilon = -ER, l = \frac{L}{\sqrt{mR}}, \quad r = \frac{mR}{L^2} \qquad (A.5)$$

we obtain the scaled energy $\varepsilon$ of the muon:

$$\varepsilon = \frac{1}{\sqrt{w^2 + v^2}} - \frac{1}{\sqrt{(1-w)^2 + v^2}} - \frac{l^2}{2v^2}. \qquad (A.6)$$

The equilibrium condition with respect to the scaled coordinate $w$ is $\partial\varepsilon/\partial w = 0$; the result can be written in the form:

$$\frac{\left((1-w)^2 + v^2\right)^{3/2}}{(w^2 + v^2)^{3/2}} = \frac{w-1}{w}. \qquad (A.7)$$

Since the left side of (A.7) is positive, the right side must also be positive: $(w-1)/w > 0$. Consequently, the allowed ranges of $w$ here are $-\infty < w < 0$ and $1 < w < +\infty$. This means that equilibrium positions of the center of the muon orbit could exist (judging only by the equilibrium with respect to $w$) either beyond the proton or beyond the electron, but there are no equilibrium positions between the proton and electron.

Solving Eq. (A.7) for $v^2$ and denoting $v^2 = p$, we obtain:

$$p(w) = w^{2/3}(w-1)^{2/3}(w^{2/3} + (w-1)^{2/3}). \qquad (A.8)$$

The equilibrium condition with respect to the scaled coordinate $v$ is $\partial\varepsilon/\partial v = 0$, which yields

$$l^2 = p^2 \left( \frac{1}{(w^2 + p)^{3/2}} - \frac{1}{((1-w)^2 + p)^{3/2}} \right). \qquad (A.9)$$

Since the left side of (A.9) is positive, the right side must be also positive. This entails the relation $w^2 + p < (1-w)^2 + p$, which simplifies to $2w - 1 < 0$, thus requiring $w < 1/2$.

So, the equilibrium with respect to both $w$ and $v$ is possible only in the range $-\infty < w < 0$, while in the second range, $1 < w < +\infty$ (derived from the equilibrium with respect to $w$ only) there is no equilibrium with respect to $v$.

From the last two relations in Eq. (A.5), we find $r = 1/\ell^2$, so that

$$r = p^{-2}\left(\frac{1}{(w^2 + p)^{3/2}} - \frac{1}{((1-w)^2 + p)^{3/2}}\right)^{-1}. \qquad (A.10)$$

where $p$ is given by (A.8). Therefore, the quantity $r$ in (A.10) is the scaled "internuclear" distance dependent on the scaled internuclear coordinate $w$.

Now we substitute the value of $\ell$ from Eq. (A.9), as well as the value of $p$ from Eq. (A.8), into Eq. (5.6) and obtain $\varepsilon(w)$ — the scaled energy of the muon dependent on the scaled internuclear coordinate $w$. Since $E = -\varepsilon/R$ and $R = rL^2/m$, then $E = -(m/L^2)\varepsilon_1$ where $\varepsilon_1 = \varepsilon/r$. The parametric dependence $\varepsilon_1(r)$ yields the energy terms.

The form of the parametric dependence $\varepsilon_1(r)$ can be significantly simplified by introducing a new parameter $\gamma = (1 - 1/w)^{1/3}$, as was shown in Chapter 3, Sec. 3. The region $-\infty < w < 0$ corresponds to $1 < \gamma < \infty$. The parametric dependence $\varepsilon_1(r)$ will then have the following form:

$$\varepsilon_1(\gamma) = \frac{(1-\gamma)^4(1+\gamma^2)^2}{2(1-\gamma+\gamma^2)^2(1+\gamma^2+\gamma^4)}, \qquad (A.11)$$

$$r(\gamma) = \frac{(1+\gamma^2+\gamma^4)^{\frac{3}{2}}}{\gamma(1+\gamma^2)^2} \qquad (A.12)$$

Classical energy terms described by the parametric dependence of the scaled energy $\varepsilon_1 = (L^2/m)E$ on the scaled internuclear distance $r = (m/L^2)R$ are presented in Fig. A.2.

Figure A.2 actually contains two coinciding energy terms: There is a double degeneracy with respect to the sign of the projection of the muon angular momentum on the internuclear axis. We remind the readers that $L$ is the absolute value of this projection — in accordance to its definition in Eq. (A.2).

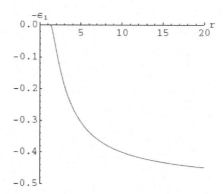

Fig. A.2. Classical energy terms: The scaled energy $-\varepsilon_1 = (L^2/m)E$ versus the scaled internuclear distance $r = (m/L^2)R$.

The minimum value of $R$, corresponding to the point where the term starts, can be found from Eq. (A.12). The term starts at $w = -\infty$, which corresponds to $\gamma = 1$; taking the value of the right side of Eq. (A.12) at this point, we find

$$R_{min} = \frac{3^{3/2}}{4} \frac{L^2}{m}. \tag{A.13}$$

With the value of $m = 206.7682746$, Eq. (A.13) yields $R = 0.00628258 \, L^2$.

The revolution frequency of the muon $\Omega$ is

$$\Omega = \frac{L}{m\rho^2} = \frac{L}{mR^2v^2} = \frac{L}{mR^2p} \tag{A.14}$$

in accordance with the previously introduced notation $p = v^2 = (\rho/R)^2$. Since $R = L^2 r/m$ (see Eq. (A.5)), then Eq. (A.14) becomes $\Omega = (m/L^3)f$, where $f = 1/(pr^2)$. Using Eq. (A.12) for $r(\gamma)$ and Eq. (A.8) for $p(w)$ with the substitution $w = 1/(1 - \gamma^3)$, where $\gamma > 1$, we finally obtain:

$$\Omega = \frac{m}{L^3}f(\gamma), \quad f(\gamma) = \frac{(1+\gamma^2)^3(1-\gamma^3)^2}{(1+\gamma^2+\gamma^4)^3}, \tag{A.15}$$

where $f(\gamma)$ is the scaled muon revolution frequency. Figure A.3 shows the scaled muon revolution frequency $f = (L^3/m)\omega$ versus the scaled internuclear distance $r = (m/L^2)R$.

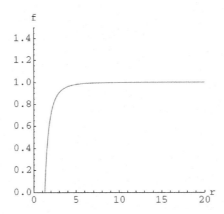

Fig. A.3. The scaled muon revolution frequency $f = (L^3/m)\Omega$ versus the scaled internuclear distance $r = (m/L^2)R$.

It is seen that for almost all values of the scaled internuclear distance $r = (m/L^2)R$, the scaled muon revolution frequency $f = (L^3/m)\Omega$ is very close to its maximum value $f_{\max} = 1$, corresponding to large values of $R$. (The quantity $f_{\max}$ can be easily found from Eq. (A.15) given that large values of $R$ correspond to $\gamma \gg 1$ and that this limit yields $f_{\max} = 1$.) In other words, for almost all values of $R$, the muon revolution frequency $\Omega$ is very close to its maximum value

$$\Omega_{\max} = \frac{m}{L^3}. \tag{A.16}$$

In the next section, we will compare the muon revolution frequency with the corresponding frequency of the electronic motion and derive the condition of validity of the separation into rapid and slow subsystems.

To analyze the stability of the muon motion, corresponding to the degenerate classical energy terms, while considering a classical circular motion of a charged particle (which was the electron in Sec. 3.2) in the field of two stationary Coulomb centers, we use the same notation as in Sec. 3.2. There it was shown that the frequencies of small oscillations of the scaled coordinates $w$ and $v$ of the circular orbit around its equilibrium position are given by

$$\omega_{\pm} = \frac{1}{(w^2 + p)^{3/4}} \sqrt{\frac{1}{1 - w} \pm \frac{3w}{Q}}, \tag{A.17}$$

where

$$Q = \sqrt{(w^2 + p)((1 - w)^2 + p)}. \tag{A.18}$$

These oscillations are in the directions $(w', v')$ obtained by rotating the $(w, v)$ coordinates by the angle $\alpha$:

$$\delta w' = \delta w \cos \alpha + \delta v \sin \alpha \quad \delta v' = -\delta w \sin \alpha + \delta v \cos \alpha \tag{A.19}$$

where the "$\delta$" symbol stands for the small deviation from equilibrium. The angle $\alpha$ is determined by the following relation:

$$\alpha = \frac{1}{2} \text{arctg} \frac{(1 - 2w)\sqrt{p}}{w(1 - w) + p}. \tag{A.20}$$

The quantity $Q$ in Eq. (A.18) is always positive since it contains the squares of the coordinates. From Eq. (A.17) it is seen that the condition for both frequencies to be real is

$$\frac{1}{1 - w} \geq \frac{3w}{Q} \tag{A.21}$$

For the frequency $\omega_-$ to be real, Eq. (A.17) requires $Q \geq 3w(1 - w)$. For any $w < 0$ (which is the allowed range of $w$), this inequality is satisfied: The left-hand side is always positive while the right-hand side is always negative.

For the frequency $\omega_+$ to be real, the following function $F(w)$ must be positive (in accordance to Eqs. (A.17) and (A.18)):

$$F(w) = (w^2 + p)((1 - w)^2 + p) - 9w^2(1 - w)^2. \tag{A.22}$$

After replacing $w$ by $\gamma = (1 - 1/w)^{1/3}$, Eq. (A.22) becomes

$$F(\gamma) = \frac{\gamma^2(\gamma^2 - 1)^2(1 + 4\gamma^2 + \gamma^4)}{(\gamma^3 - 1)^4}. \tag{A.23}$$

Since the allowed range of $w < 0$ corresponds to $\gamma > 1$, it is seen that $F(\gamma)$ is always positive.

Thus, the corresponding classical energy terms represent the stable motion.

Now we unfreeze the slow subsystem and analyze a slow revolution of the axis connecting the proton and electron, the electron executing a circular orbit. In accordance with the concept of separating rapid and slow subsystems, the rapid subsystem (the revolving muon) follows the adiabatic evolution of the slow subsystem. This means that the slow subsystem can be treated as a modified "rigid rotator" consisting of the electron, the proton, and the ring, over which the muon charge is uniformly distributed, all distances within the system being fixed (see Fig. A.1).

The potential energy of the electron in atomic units (with the angular-momentum term) is

$$E_e = \frac{M^2}{2R^2} - \frac{1}{R} + \frac{1}{\sqrt{\rho^2 + (R - z)^2}}, \tag{A.24}$$

where $M$ is the electron angular momentum. Its derivative by $R$ must vanish at equilibrium, which yields

$$\frac{dE_e}{dR} = -\frac{M^2}{R^3} + \frac{1}{R^2} - \frac{R - z}{\left(\rho^2 + (R - z)^2\right)^{3/2}} = 0. \tag{A.25}$$

This gives us the value of the scaled angular momentum

$$l_e = \frac{M}{\sqrt{R}} \tag{A.26}$$

corresponding to the equilibrium:

$$l_e^2 = 1 - \frac{1 - w}{\left((1 - w)^2 + p\right)^{3/2}}, \tag{A.27}$$

where the scaled quantities $w$, $p$ of the muon coordinates are defined in Eq. (A.5). Using the muon equilibrium condition from Eq. (A.7) with $v^2$ denoted as $p$, we can represent Eq. (A.27) in the form

$$l_e^2 = 1 + \frac{w}{(w^2 + p)^{3/2}}. \tag{A.28}$$

After replacing $w$ by $\gamma = (1 - 1/w)^{1/3}$, we obtain

$$l_e(\gamma) = \sqrt{1 - \frac{(1 - \gamma)^2\sqrt{1 + \gamma + \gamma^2}}{(1 - \gamma + \gamma^2)^{\frac{3}{2}}}}. \tag{A.29}$$

The electron revolution frequency is $\omega = M/R^2 = l_e(\gamma)/R^{3/2}$ given that $M = l_e(\gamma)R^{1/2}$ in accordance to Eq. (A.26). Since $R =$

$L^2 r(\gamma)/m$ (see Eq. (A.5)) with $r(\gamma)$ given by Eq. (A.12), then from $\omega = \ell_e(\gamma)/R^{3/2}$ we obtain

$$\omega = \frac{m^{3/2} l_e(\gamma)}{L^3 r^{3/2}(\gamma)}. \qquad (A.30)$$

From Eqs. (A.15) and (A.30) we find the following ratio of the muon and electron revolution frequencies:

$$\frac{\Omega}{\omega} = \frac{1}{\sqrt{m}} \frac{f(\gamma) r^{3/2}(\gamma)}{l_e(\gamma)}, \qquad (A.31)$$

where $f(\gamma)$ is given in Eq. (A.15).

In addition to the above relation $R = L^2 r(\gamma)/m$, the same quantity $R$ can be expressed from Eq. (A.26) as $R = M^2/\ell_e^2(\gamma)$. Equating the right-hand sides of these two expressions, we obtain the equality $L^2 r(\gamma)/m = M^2/\ell_e^2(\gamma)$, from which it follows:

$$\frac{L}{M} = \frac{\sqrt{m}}{l_e(\gamma)\sqrt{r(\gamma)}}. \qquad (A.32)$$

The combination of Eqs. (A.31) and (A.32) represents an analytical dependence of the ratio of the muon and electron revolution frequencies $\Omega/\omega$ versus the ratio of the muon and electron angular momenta $L/M$ via the parameter $\gamma$ as the latter varies from 1 to $\infty$. This dependence is presented in Fig. A.4.

For the separation into the rapid and slow subsystems to be valid, the ratio of frequencies $\Omega/\omega$ should be significantly greater than unity. From Fig. A.4 it is seen that this requires the ratio of angular momenta $L/M$ to be noticeably greater than 20.

There is another validity condition to be checked for this scenario. Namely, the revolution frequency $\Omega$ of the muon must also be much greater than the inverse lifetime of the muon $1/T_{\text{life}}$, where $T_{\text{life}} = 2.2\,\mu s = 0.91 \times 10^{11}$ a.u., namely: $\Omega T_{\text{life}} \gg 1$. Since for almost all values of $R$, the muon revolution frequency $\Omega$ is very close to its maximum value $\Omega_{\max} = m/L^3$, as shown above, then the second validity condition can be estimated as $(m/L^3)T_{\text{life}} \gg 1$, from which it follows

$$L \ll L_{max} = (mT_{life})^{\frac{1}{3}} = 26,600 \qquad (A.33)$$

(we remind that $m = 206.7682746$ in atomic units). So, the second validity condition is fulfilled for any practically feasible value of the muon angular momentum $L$.

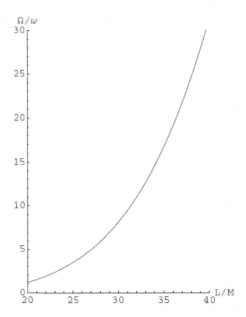

Fig. A.4. The ratio of the muon and electron revolution frequencies $\Omega/\omega$ versus the ratio of the muon and electron angular momenta $L/M$.

Thus, for the ratio of angular momenta $L/M$ noticeably greater than 20, we deal here with a muonic quasimolecule, where the muon rapidly rotates about the axis connecting the proton and electron, while the plane of the muon orbit follows a relatively slow rotation of this axis.

The muon, rotating in a circular orbit at the frequency $\Omega(R)$, should emit a spectral line at this frequency. The maximum value $\Omega_{\max} = m/L^3$ corresponds to the frequency of spectral lines emitted by the muonic hydrogen atom (by the $\mu p$-subsystem). For the equilibrium value of the proton–electron separation — just as for almost all values of $R$ — the frequency $\Omega$ is slightly smaller than $\Omega_{\max}$. Therefore, the spectral lines, emitted by the muon in the quasimolecule $\mu pe$, experience a red shift compared to the corresponding spectral lines that would have been emitted by the muon in a muonic hydrogen atom. The relative red shift $\delta$ is defined as follows

$$\delta = \frac{\lambda - \lambda_0}{\lambda_0} = \frac{\Omega_{\max} - \Omega}{\Omega}, \tag{A.34}$$

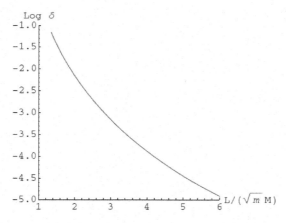

Fig. A.5. Universal dependence of the relative red shift $\delta$ of the spectral lines of the quasimolecule $\mu pe$ (or $\pi pe$) on $L/(m^{1/2}M)$, which is the ratio of the muon and electron angular momenta $L/M$ divided by the square root of the mass $m$ of the muon or pion.

where $\lambda$ and $\lambda_0$ are the wavelength of the spectral lines for the quasimolecule $\mu pe$ and the muonic hydrogen atom, respectively. Using Eq. (A.15), the relative red shift can be represented in the form

$$\delta(\gamma) = \frac{1}{f(\gamma)} - 1, \tag{A.35}$$

where $f(\gamma)$ is given in Eq. (A.15).

The combination of Eqs. (A.35) and (A.32) represents an analytical dependence of the relative red shift $\delta$ on the ratio of the muon and electron angular momenta $L/M$ via the parameter $\gamma$ as the latter varies from 1 to $\infty$. Figure A.5 presents the dependence of $\delta$ on $L/(m^{1/2}M)$. In this form the dependence is "universal", i.e., valid for different values of the mass $m$: For example, it is valid also for the quasimolecule $\pi pe$ where there is a pion instead of the muon. Figure A.6 presents the dependence of $\delta$ on $L/M$ specifically for the quasimolecule $\mu pe$.

It is seen that is the relative red shift of the spectral lines is well within the spectral resolution $\Delta\lambda_{\mathrm{res}}/\lambda$ of available spectrometers: $\Delta\lambda_{\mathrm{res}}/\lambda \sim (10^{-4}-10^{-5})$ as long as the ratio of the muon and electron angular momenta $L/M < 80$. Thus, this red shift can be observed and this would be one of the ways to detect the formation of such muonic negative hydrogen ions.

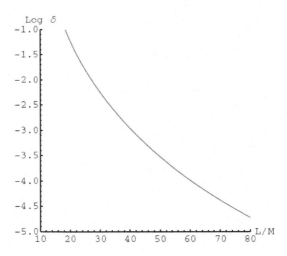

Fig. A.6. Dependence of the relative red shift $\delta$ of the spectral lines of the quasimolecule $\mu pe$ on the ratio of the muon and electron angular momenta $L/M$.

Figure A.7 presents the dependence of the relative red shift $\delta$ on the ratio of the muon and electron revolution frequencies $\Omega/\omega$. It is seen that the relative red shift decreases as the ratio of the muon and electron revolution frequencies increases, but it remains well within the spectral resolution $\Delta\lambda_{\text{res}}/\lambda$ of available spectrometers.

Finally it should be noted that in paper [A.12] we generalized the above study by replacing the proton in $\mu pe$ quasimolecule by a fully-stripped ion of a nuclear charge $Z > 1$. We showed that in this case, just as in the above case of $Z = 1$, the muonic motion can represent a rapid subsystem while the electronic motion can represent a slow subsystem. For this to be valid, the ratio of the muonic and electronic angular momenta should be slightly greater than in the case of $Z = 1$. We demonstrated that the binding energies of the muon for $Z > 1$ are much greater than for $Z = 1$ at any finite value of the nucleus–electron distance. Finally we showed that the red shift of the spectral lines emitted by the muon (compared to the spectral lines of the corresponding muonic hydrogen-like ion of the nuclear charge $Z$) decreases as $Z$ increases. However, the relative red shift remains within the spectral resolution of available spectrometers at

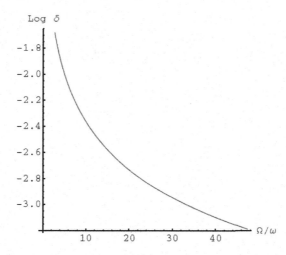

Fig. A.7. Dependence of the relative red shift $\delta$ on the ratio of the muon and electron revolution frequencies $\Omega/\omega$.

least up to $Z = 5$. Observing this red shift should be one of the ways to detect the formation of the quasimolecules $\mu Ze$.

## Appendix B

# Helical and Circular States of Diatomic Rydberg Quasimolecules in a Laser Field

In Chapter 3 we presented classical analytical studies of Rydberg states of two-Coulomb-center systems consisting of two nuclei of charges $Z$ and $Z'$, separated by a distance $R$, and one electron. Energy terms of these Rydberg quasimolecules were obtained for a field-free case, as well as under a static electric field, or under a static magnetic field [1.20], or under the screening by plasma electrons.

Here we present the classical analytical study of such Rydberg quasimolecules under a laser field. For the situation where the laser field is *linearly-polarized* along the internuclear axis, we find an analytical solution for the stable helical motion of the electron, valid for wide ranges of the laser field strength and frequency. We also found resonances, corresponding to a laser-induced unstable motion of the electron, resulting in the destruction of the helical states. For the situation where such Rydberg quasimolecules are under a *circularly-polarized* laser field, the polarization plane being perpendicular to the internuclear axis, we find an analytical solution for circular Rydberg states, valid for wide ranges of the laser field strength and frequency.

We show that both under the linearly-polarized laser field and under the circularly-polarized laser field, in the electron radiation spectrum, in the addition to the primary spectral component at (or near) the unperturbed revolution frequency of the electron, there appear *satellites*. We find that for the case of the linearly-polarized laser field, the intensities of the satellites are proportional to the squares of the Bessel functions $J_q^2(s)$, $(q = 1, 2, 3, \ldots)$, where $s$ is proportional to the laser field strength. As for the case of the circularly-polarized field, we demonstrate that there is a red shift of the primary spectral component — the shift linearly proportional to the laser field strength.

First we consider the case where the laser is polarized parallel to the internuclear axis and oscillates sinusoidally with the frequency $\omega$. The angular momentum $M$ is conserved here due to $\varphi$-symmetry. The corresponding Hamiltonian is

$$H = \frac{p_\rho^2 + p_\varphi^2}{2} + \frac{M^2}{2\rho^2} - \frac{Z}{\sqrt{\rho^2 + z^2}} - \frac{Z'}{\sqrt{\rho^2 + (R - z)^2}} + zF\cos\omega t.$$

(B.1)

Below we scale all frequencies using the factor $(R^3/Z)^{1/2}$: For example, the scaled laser frequency is $\mu = \omega(R^3/Z)^{1/2}$. We also use scaled coordinates

$$w = \frac{z}{R}, \quad v = \frac{\rho}{R},$$

(B.2)

where $R$ is the internuclear distance. The origin is at the location of charge $Z$.

Without the electric field, in the vicinity of the equilibrium the motion in $z\rho$-space corresponds to a two-dimensional harmonic oscillator, as shown in Chapter 3, Sec. 2. Its scaled eigen-frequencies are

$$\omega_\pm = \frac{1}{(w^2 + v^2)^{3/4}} \sqrt{\frac{1}{1 - w} \pm \frac{3w}{\sqrt{(w^2 + v^2)((1 - w)^2 + v^2)}}},$$

(B.3)

where the equilibrium value of $v$ connected to $w$ as follows:

$$v(w, b) = \sqrt{\frac{w^{\frac{2}{3}}(1 - w)^{\frac{4}{3}} - b^{\frac{2}{3}}w^2}{b^{\frac{2}{3}} - w^{\frac{2}{3}}(1 - w)^{-\frac{2}{3}}}}.$$

(B.4)

The motion occurs on the axes $(w', v')$ — they are the original axes $(w, v)$ rotated by some angle $\alpha$. The dependence of the angle $\alpha$ on the scaled coordinate $w$ can be expressed in the most compact form by introducing the notation:

$$\gamma = \left(\frac{1}{w} - 1\right)^{1/3}.$$ (B.5)

In the $\gamma$-representation the angle $\alpha$ has the form

$$\alpha = \frac{1}{2}\arctan\frac{\sqrt{(b^{2/3}\gamma^2 - 1)(\gamma^4 - b^{2/3})}}{\gamma(b^{2/3} + \gamma)}.$$ (B.6)

The scaled eigen-frequencies $\omega_-$ and $\omega_+$ are the scaled frequencies of small oscillations about the equilibrium along the coordinates $w'$ and $v'$, respectively.

As we introduce the oscillating electric field, these oscillations become forced, with the forces $F\cos\alpha \cos\omega t$ on $w'$ and $F\sin\alpha \cos\omega t$ on $v'$. Therefore, the deviations from equilibrium on $(w', v')$ are (see, e.g., textbooks [B.1, B.2])

$$\delta w' = \frac{f\cos\alpha}{\omega_-^2 - \mu^2}\cos\mu\tau, \quad \delta v' = \frac{f\sin\alpha}{\omega_+^2 - \mu^2}\cos\mu\tau$$ (B.7)

where $\mu = \omega(R^3/Z)^{1/2}$ and $\tau = t(Z/R^3)^{1/2}$. Now we revert to the original coordinates $(w, v)$ and obtain the equations of motion in the linearly-polarized oscillatory electric field in the vicinity of the equilibrium: The electron follows the circular path corresponding to the case with no electric field with the following deviations from equilibrium depending on the scaled time $\tau$:

$$\delta w = f\left(\frac{\cos^2\alpha}{\omega_-^2 - \mu^2} + \frac{\sin^2\alpha}{\omega_+^2 - \mu^2}\right)\cos\mu\tau,$$

$$\delta v = f\sin\alpha\cos\alpha\left(\frac{1}{\omega_-^2 - \mu^2} - \frac{1}{\omega_+^2 - \mu^2}\right)\cos\mu\tau.$$ (B.8)

From Eq. (B.8) it is seen that the strength and frequency of the laser field affect the amplitudes of the forced oscillations on $w$- and $v$-axes. In fact, these amplitudes are proportional to the field strength $f$. The frequencies of the forced oscillations on the axes are equal to that of the laser field, instead of $\omega_-$ and $\omega_+$.

Since the Hamiltonian from Eq. (B.1) does not depend on $\varphi$, the corresponding momentum is conserved:

$$p_\varphi = \rho^2 \frac{d\varphi}{dt} \equiv M = \text{const.} \tag{B.9}$$

We can rewrite Eq. (B.9) in the scaled notation as

$$\frac{d\varphi}{d\tau} = \frac{l}{v^2(\tau)}, \tag{B.10}$$

where $\ell = M/(ZR)^{1/2}$ is the scaled angular momentum. Substituting in Eq. (B.10) $v(\tau) = v_0 + \delta v(\tau)$, where $v_0(w)$ is the equilibrium value of the scaled radius $v$ of the electron orbit from (4.4) and $\delta v(\tau)$ is given by Eq. (B.8), we obtain

$$\frac{d\varphi}{dt} \approx \frac{l}{v_0^2} - \frac{2l}{v_0^3} \delta v(\tau) \tag{B.11}$$

which after the integration with respect to time yields:

$$\varphi(t) \approx \frac{l}{v_0^2}\tau - \frac{2l}{\mu v_0^3} f \sin\alpha\cos\alpha \left( \frac{1}{\omega_-^2 - \mu^2} - \frac{1}{\omega_+^2 - \mu^2} \right) \sin\mu\tau. \tag{B.12}$$

From Eq. (B.12) it is seen that the $\varphi$-motion is a rotation about the internuclear axis with the scaled frequency $\ell/v_0^2$, slightly modulated by oscillations of the scaled radius of the orbit $v$ at the laser frequency $\mu$ (i.e., at the laser frequency $\omega$ in the usual notation).

Thus, from Eqs. (B.8) and (B.12) it is clear that the electron is bound to a *conical surface* which incorporates the original circular orbit. In Fig. B.1, the three-dimensional actual trajectory is plotted for $b = 3$, $f = 1$, $\mu = 1$ at $w = 0.2$.

The expression for $\varphi(\tau)$ from Eq. (B.12), i.e., $\varphi(t(Z/R^3)^{1/2})$, enters the following Fourier-transform that determines the amplitude of the power spectrum of the electron radiation

$$A_l(\Delta) = \frac{1}{\pi} dt \cos\left( \Delta t - \varphi\left( t\sqrt{\frac{z}{R^3}} \right) \right), \tag{B.13}$$

where $\Delta$ is the radiation frequency measured, e.g., by a spectrometer. The sinusoidal modulation of the phase $\varphi$ is analogous to the situation where hydrogen spectral lines are modified by an external

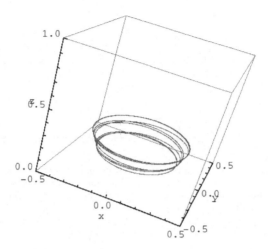

Fig. B.1. The actual trajectory of the electron in the linearly-polarized laser field for $b = 3$, $f = 1$, $\mu = 1$ at $w = 0.2$. The $z$-axis is along the internuclear axis.

monochromatic field at the frequency $\omega$, the latter problem being solved analytically by Blochinzew as early as in 1933 [B.3] (a further study can be found, e.g., in book [B.4]).

From Blochinzew's results it follows for our case that in the electron radiation spectrum, the helical motion should manifest as follows. The most intense emission would be at the frequency $\Omega = d\varphi/dt$ of the rapid $\varphi$-motion. In addition, there will be satellites at the frequencies $\Omega \pm q\omega$, where $q = 1, 2, 3, \ldots$, whose relative intensities $I_q$ are controlled by the Bessel functions $J_q(s)$:

$$I_q = J_q^2(s), \quad s = \frac{2l}{\mu v_0^3} f \sin\alpha \cos\alpha \left( \frac{1}{\omega_-^2 - \mu^2} - \frac{1}{\omega_+^2 - \mu^2} \right). \quad (\text{B.14})$$

The oscillatory motion of the electron in the $z\rho$-space with the laser frequency $\omega$ should lead also to the radiation at this frequency. However, since $\omega \ll \Omega$, this spectral component would be far away from the primary spectral line and its satellites.

From Eq. (B.8) it is also seen that there are resonances when the laser frequency is equal to one of the eigen-frequencies of the motion in the $z\rho$-space, i.e., when either $\mu = \omega_+$ or $\mu = \omega_-$. It turns out that these conditions yield three resonance points on the $w$-axis for

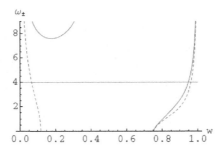

Fig. B.2. Eigen-frequencies of the motion in the $z\rho$-space $\omega_+$ (solid curves) and $\omega_-$ (dashed curves) versus $w$, i.e., versus the scaled $z$-coordinate of original circular Rydberg state. The scaled laser frequency $\mu$ is shown by the horizontal straight line. The plot is for $b = 3$ and $\mu = 4$. Three resonant points are seen.

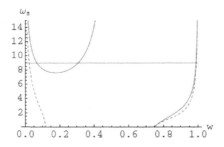

Fig. B.3. Same as in Fig. B.2, but for $b = 3$ and $\mu = 9$. Five resonant points are seen.

the laser field frequency $\mu$ below a certain critical value $\mu_c$, or five resonance points for $\mu > \mu_c$ , as shown in Figs. (B.2) and (B.3).

For instance, in the case of $b = 3$, for $\mu = 8$, we observe resonances at the following five values of $w$: 0.02883, 0.1106, 0.2497, 0.9852, 0.9878. The critical value corresponds to the minimum of $\omega_+(w)$ for a given $b$ in the interval $0 < w < w_1$ at the equilibrium point (the equilibrium scaled coordinate $v$ being expressed via $w$ by (B.4)). Calculating the derivative of $\omega_+$ with respect to $w$ and setting it equal to zero, we find the point of the minimum. The value of $\omega_+$ at this point will be equal to the critical value of the scaled laser frequency $\mu_c$. For example, for $b = 3$ at $w = 0.17642$ (the minimum of $\omega_+$ in Figs. B.2 and B.3) this critical value is $\mu_c = 7.5944$. As the

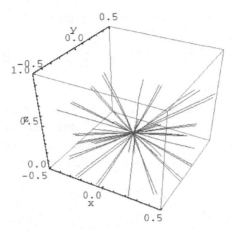

Fig. B.4. The actual trajectory of the electron (for various directions of its initial velocity) in the linearly-polarized laser field for a resonance case where $b = 3$, $f = 1$, $\mu = 8$, and $w = 0.111$. The $z$-axis is along the internuclear axis.

ratio of nuclear charges $b$ increases, so does also the critical value $\mu_c$ of the scaled laser frequency.

These resonances correspond to a laser-induced unstable motion of the electron, resulting in the destruction of the helical states. This is illustrated in Fig. B.4 showing the three-dimensional actual trajectory of the electron (for various directions of its initial velocity) for a resonance case where $b = 3$, $f = 1$, $\mu = 8$, and $w = 0.111$ ($w = 0.111$ is one of the three values of $w$, at which the laser frequency $\mu$ coincides with the eigen-frequency $\omega_+$). A striking difference is seen compared to the stable helical motion depicted in Fig. B.1: The resonance destroyed the helical state.

Now we consider the case of a circular polarization of the laser field, polarization plane being perpendicular to the internuclear axis. The laser field varies as

$$\mathbf{F} = F(\mathbf{e}_x \cos\omega t + \mathbf{e}_y \sin\omega t) \tag{B.15}$$

where $\mathbf{e_x}$ and $\mathbf{e_y}$ are the unit vectors along the $x$- and $y$-axes, $F$ is the amplitude and $\omega$ is the frequency. The Hamiltonian for the electron in this configuration will take the following form:

$$H = \frac{1}{2}\left(p_\rho^2 + p_z^2 + \frac{p_\varphi^2}{\rho^2}\right) - \frac{Z}{\sqrt{\rho^2 + z^2}} - \frac{Z'}{\sqrt{\rho^2 + (R-z)^2}}$$
$$+ F\rho\cos(\varphi - \varphi_0), \tag{B.16}$$

where we introduced $\varphi_0 = \omega t$. Again, we consider $\varphi$-motion to be the rapid subsystem, i.e., $d\varphi/dt$ is much greater than the laser frequency $\omega$ and the frequencies of $z$- and $\rho$-motion. The canonical equations for the $\varphi$-motion obtained from Eq. (B.16) are

$$\frac{d\varphi}{dt} = \frac{\partial H}{\partial p_\varphi} = \frac{p_\varphi}{\rho^2}, \tag{B.17}$$

$$\frac{dp_\varphi}{dt} = -\frac{\partial H}{\partial \varphi} = F\rho\sin(\varphi - \varphi_0) \tag{B.18}$$

Combining Eqs. (B.17) and (B.18), we get

$$\frac{d^2\varphi}{dt^2} = \frac{F}{\rho}\sin(\varphi - \varphi_0). \tag{B.19}$$

After a substitution $\varphi - \varphi_0 = \theta + \pi$, Eq. (B.19) becomes

$$\frac{d^2\theta}{dt^2} = -\frac{F}{\rho}\sin\theta, \tag{B.20}$$

which is the equation of motion of a mathematical pendulum of length $\rho$ in gravity $F$. Its two possible modes are liberation and rotation; since $\theta$ is the rapid coordinate, we have the case of rotation. The solution for $\theta(t)$ is well-known and can be expressed in terms of the special function known as Jacobi amplitude:

$$\theta(t) = 2\,\mathrm{am}\left(\frac{\Omega t}{2}, \frac{4F}{\rho\Omega^2}\right). \tag{B.21}$$

Here we denoted $d\theta/dt$ at $t = 0$ as $\Omega$. For rapid rotations, the change in the angular speed on $\theta$ is insignificant compared to the initial speed and $d\theta/dt \approx \Omega$.

The expression for $\theta(t)$ enters the following Fourier-transform that determines the amplitude of the power spectrum of the electron radiation:

$$A_c\left(\Delta, \frac{4F}{\rho\Omega^2}\right) = \frac{1}{\pi}\int_0^\infty dt\cos\left(\Delta t - \theta\left(t, \frac{4F}{\rho\Omega^2}\right)\right). \tag{B.22}$$

Figure B.5 shows as an example the power spectrum of the electron radiation spectrum (i.e., $A_c^2$) versus the dimensionless radiation

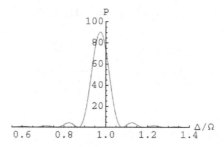

Fig. B.5. The power spectrum of the electron radiation $P$ (in arbitrary units) versus the dimensionless radiation frequency $\Delta/\Omega$ for the case where $4F/(\rho\Omega) = 0.1$. Here $\Omega$ is the frequency of the electron radiation at the absence of the laser field. A certain width is assigned to all spectral components to display a continuous spectral line profile.

frequency $\Delta/\Omega$ for the case where $4F/(\rho\Omega) = 0.1$. It is seen that the most intense component in the spectrum is at the frequency $\Delta$ approximately equal to, but slightly less than $\Omega$. It is also seen that the laser modulation of the primary frequency of the electron rotation results in a series of relatively small satellites of the primary spectral component.

The red shift of the primary spectral component can be calculated analytically as follows. Since $\varphi$-motion is rapid, we can average the Hamiltonian in Eq. (B.16) with respect to time. Integrating Eq. (B.20) with the initial condition $d\theta/dt = \Omega$, we get

$$\Omega^2 - \left(\frac{d\theta}{dt}\right)^2 = \frac{4F}{\rho}\sin^2\frac{\theta}{2}. \tag{B.23}$$

By averaging this equation with respect to time, we obtain

$$\Omega^2 - \left\langle\frac{d\theta^2}{dt}\right\rangle = \frac{2F}{\rho}. \tag{B.24}$$

Thus, the $\varphi$-momentum term in the Hamiltonian (B.16) becomes

$$\left\langle\frac{p_\varphi^2}{\rho^2}\right\rangle = \rho^2\left\langle\left(\frac{d\theta}{dt}\right)^2\right\rangle = \rho^2\Omega^2\left(1 - \frac{2F}{p\Omega^2}\right). \tag{B.25}$$

The last term in the Hamiltonian from Eq. (B.16) vanishes after the time averaging so that the time-averaged Hamiltonian depends only on $\rho$- and $z$-coordinates and their corresponding momenta. The result is the following quasi-stationary Hamiltonian with no explicit time dependence:

$$H = \frac{1}{2}(p_\rho^2 + p_z^2) - \frac{Z}{\sqrt{\rho^2 + z^2}} - \frac{Z'}{\sqrt{\rho^2 + (R - z)^2}} + \frac{1}{2}\rho^2\Omega^2 - \rho F.$$

(B.26)

Introducing the scaled quantities

$$w = \frac{z}{R}, \quad v = \frac{\rho}{R}, \quad f = \frac{FR^2}{Z}, \quad \sigma = \Omega\sqrt{\frac{R^3}{Z}} \qquad \text{(B.27)}$$

and using the Hamilton equations, we obtain the following two differential equations of motion:

$$-\ddot{w} = \frac{w}{(w^2 + v^2)^{\frac{3}{2}}} - \frac{b(1 - w)}{((1 - w)^2 + v^2)^{\frac{3}{2}}}, \qquad \text{(B.28)}$$

$$-\ddot{v} = v\left(\frac{1}{(w^2 + v^2)^{3/2}} + \frac{b}{((1 - w^2) + v^2)^{3/2}} + \sigma^2\right) - f \quad \text{(B.29)}$$

(the dot above the letter indicates the differentiation by the scaled time $\tau = t(Z/R^3)^{1/2}$).

Here we consider these Rydberg quasimolecules in circular (not helical) states, so that the plane of the electron orbit has a stationary position on the internuclear axis. Therefore, the right side of Eq. (B.28) vanishes and the relationship between $w$ and $v$ becomes the same as given by Eq. (B.4). This makes the scaled radius of the orbit $v$ a constant as well.

Since the angular momentum is $M = \Omega\rho^2$ for a stationary circular orbit, the averaging of the $\varphi$-momentum in Eq. (B.25) is equivalent to changing $M$ to $M(1 - F\rho^3/M^2)$. Using scaled units and the relationship $M = \Omega\rho^2$, we find out that the case of the circularly-polarized laser field is equivalent to a field-free case, but with an effective frequency $\Omega$ given by the substitution:

$$\Omega \to \Omega(1 - \varkappa(\gamma)f), \qquad \text{(B.30)}$$

where

$$\varkappa(\gamma) = \frac{\gamma^6(\gamma^3 - 1)^{3/2}(\gamma^4 - b^{2/3})^{3/2}}{(\gamma^3 + 1)^{11/2}(b^{2/3}\gamma^2 - 1)^3}.$$  (B.31)

The quantity $\Omega\varkappa(\gamma)f$ is the red shift of the primary spectral component. This result is valid as long as the relative correction $\varkappa(\gamma)f$ to the unperturbed angular frequency $\Omega$ of the electron remains relatively small. Figures B.6 and B.7 illustrate the situation for the case where the ratio of the nuclear charges is $b = 2$. On the horizontal axis is the scaled coordinate $w$, i.e., the scaled coordinate along the internuclear axis of the Rydberg quasimolecule. The solid curve, having two branches, shows the unperturbed angular frequency $\Omega$ of the electron. The dashed curve shows the correction $\Omega\varkappa(\gamma)f$. It is seen that the correction remains relatively small for the entire left branch

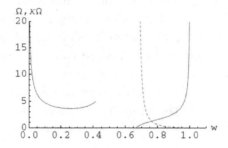

Fig. B.6. Dependence of the unperturbed angular frequency $\Omega$ of the electron (solid curve, two branches) and of the correction $\Omega\varkappa(\gamma)f$ for $f = 1$ (dashed curve) on the scaled coordinate $w$ along the internuclear axis of the Rydberg quasimolecule.

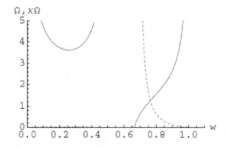

Fig. B.7. Same as in Fig. B.6, but with better visible details in the region of the right branch of $\Omega(w)$.

of $\Omega$ and for a significant part of the right branch of $\Omega$. (Figures B.6 and B.7 differ only by the range of the vertical scale, so that Fig. B.6 allows to see more clearly the region where the solid and dashed curves intersect and the region of validity of the results for the right branch of $\Omega$.) Physically, the left branch corresponds to the situation where the electron is primarily bound by the charge $Z$. The region of the right branch, where the correction is relatively small, physically corresponds to the situation where the electron is primarily bound by the charge $Z'$.

Thus, under a laser field of a known strength, in the case of the linear polarization the observation of the satellites would be the confirmation of the helical electronic motion in the Rydberg quasimolecule, while in the case of the circular polarization the observation of the red shift of the primary spectral component would be the confirmation of the specific type of the phase modulation of the electronic motion described by Eq. (B.15). Conversely, if the laser field strength is unknown, both the relative intensities of the satellites and the red shift of the primary spectral component could be used for measuring the laser field strength.

Finally we note that in quantum mechanics, spectral satellites can be best described using the formalism of quasienergy states, such as the states of the combined system "Rydberg quasimolecule + laser field" (see, e.g., [B.4]).

# References

## Chapter 1

1.1. E. Oks, *J. Phys. B: At. Mol. Opt. Phys.* **34** (2001) 2235.

1.2. B.H. Bransden and M.R.C. McDowell, *Charge Exchange and the Theory of Ion–Atom Collisions* (Oxford University Press, Oxford) 1992.

1.3. B.M. Smirnov, *Physics of Atoms and Ions* (Springer, Berlin) 2003, Sec. 14.3.

1.4. S.S. Gershtein and V.D. Krivchenkov, *Sov. Phys. JETP* **13** (1961) 1044.

1.5. E. Oks, *Phys. Rev. Lett.* **85** (2000) 2084.

1.6. E. Oks, *J. Phys. B: At. Mol. Opt. Phys.* **33** (2000) 3319.

1.7. M.R. Flannery and E. Oks, *Phys. Rev. A* **73** (2006) 013405.

1.8. E. Oks, *Stark Broadening of Hydrogen and Hydrogenlike Spectral Lines in Plasmas: The Physical Insight* (Alpha Science International, Oxford) 2006, Appendix A.

1.9. N. Kryukov and E. Oks, *J. Phys. B: At. Mol. Opt. Phys.* **46** (2013) 245701.

1.10. N. Kryukov and E. Oks, *Eur. Phys. J. D* **68** (2014) 171.

1.11. N. Kryukov and E. Oks, *Can. J. Phys.* **91** (2013) 715.

1.12. N. Kryukov and E. Oks, *Can. J. Phys.* **90** (2012) 647.

1.13. N. Kryukov and E. Oks, *Int. Rev. At. Mol. Phys.* **4**, No. 2 (2013).

1.14. L.D. Landau and E.M. Lifshitz, *The Classical Theory of Fields* (Pergamon, Oxford) 1971.

1.15. P.A.M. Dirac, *Can. J. Phys.* **2** (1950) 129.

1.16. P.A.M. Dirac, *Proc. R. Soc. A* **246** (1958) 326.

1.17. P.A.M. Dirac, *Lectures on Quantum Mechanics* (Academic, New York) 1964.

1.18. P.A.M. Dirac, *Lectures on Quantum Theory* (Academic, New York) 1966.

1.19. E. Oks, *J. Phys. B: At. Mol. Phys.* **34** (2001) 2235.

1.20. J.A. Camarena and E. Oks, *Int. Rev. At. Mol. Phys.* **1** (2010) 143.

1.21. G.V. Sholin and A.E. Trenin, *J. Exp. Theor. Phys.* **112** (2011) 913.

1.22. D.R. Inglis and E. Teller, *Astrophys. J.* **90** (1939) 439.

1.23. E. Oks, *Int. Rev. At. Mol. Phys.* **4**, No. 2 (2013).

1.24. E. Oks and E. Leboucher-Dalimier, Phys. Rev. E, Rapid Communications **62**(2000) R3067.

1.25. E. Oks, *Stark Broadening of Hydrogen and Hydrogenlike Spectral Lines in Plasmas: The Physical Insight* (Oxford: Alpha Science International) 2006, Sect. 8.2.

1.26. E. Oks, and E. Leboucher-Dalimier, *J. Phys. B: Atom Mol. Opt. Phys.* **33** (2000) 3795.

1.27. St. Boeddeker, H.-J. Kunze, and E. Oks, *Phys. Rev. Lett.* **75** (1995) 4740.

1.28. E. Leboucher-Dalimier, E. Oks, E. Dufour, P. Sauvan, P. Angelo, R. Schott, and A. Poquerusse, Phys. Rev. E, Rapid Communications **64** (2001) 065401.

1.29. E. Dalimier, E. Oks, O. Renner, and R. Schott, *J. Phys. B: Atom. Mol. Opt. Phys.* **40** (2007) 909.

1.30. O. Renner, E. Dalimier, R. Liska, E. Oks, and M. Šmid, *J. Phys. Conf. Ser.* **397** (2012) 012017.

1.31. E. Dalimier and E. Oks, *J. Phys. B: Atom. Mol. Opt. Phys.* **47** (2014) 105001.

1.32. E. Dalimier, E. Oks, and O. Renner, Atoms **2** (2014) 178.

## Chapter 2

2.1. V. Fock, *Z. Phys.* **98** (1935) 145.

2.2. M. Gryzinski, *Phys. Rev. Lett.* **14** (1965) 1059.

2.3. M. Gryzinski, *Phys. Lett.* **41A** (1972) 69.

2.4. P. Grujuc, A. Tomic and S. Vucic, *J. Chem. Phys.* **79** (1983) 1776.

2.5. R.A. Mapleton, *Proc. Phys. Soc.* **87** (1966) 219.

2.6. A.E. Kingston, *Phys. Rev.* **135** (1964) A1537; *Proc. Phys. Soc.* **87** (1966) 193.

2.7. M. Gryzinski, *Proc. 6th Int. Conf. on Ionization Phenomena in Gases* (S.E.R.M.A., Paris) 1963, p. 117.

2.8. M. Gryzinski, *Phys. Rev.* **138** (1965) A336.

2.9. I.C. Percival and D. Richards, *J. Phys. B: At. Mol. Phys.* **3** (1970) 315.

2.10. M.R. Flannery and D. Vrinceanu, *Phys. Rev. Lett.* **85** (1970) 1.

2.11. L. Vriens, in *Case Studies in Atomic Collision Physics 1*, eds. E.W. McDaniel and M.R.C. McDowell (North-Holland, Amsterdam) 1969, p. 335.

2.12. A. Burgess and I.C. Percival, in *Advances in Atomic and Molecular Physics*, eds. D.R. Bates and I. Estermann,**v. 4** (Academic, New York) 1968, p. 109.

2.13. D.R. Bates and A.E. Kingston, in *Advances in Atomic and Molecular Physics*, eds. D.R. Bates and I. Estermann,**v. 6** (Academic, New York) 1970, p. 269.

2.14. M. Gryzinski, in *Classical Dynamics in Atomic and Molecular Physics*, eds. T. Grozdanov, P. Grujic and P. Krstic (World Scientific, Singapore) 1989, p. 50.

2.15. M. Gryzinski and J.A. Kunc, *J. Phys. B: At. Mol. Phys.* **32** (1999) 5789.

2.16. R. Abrines and I.C. Percival, *Proc. Phys. Soc.* **88** (1966) 861.

2.17. R. Abrines, I.C. Percival and N.A. Valentine, *Proc. Phys. Soc.* **89** (1966) 515.

2.18. M. Gryzinski, J. Kunc and M. Zgorzelski, *Phys. Lett.* **38A** (1972) 35.

2.19. M. Gryzinski, J. Kunc and M. Zgorzelski, *J. Phys. B: At. Mol. Phys.* **6** (1973) 2292.

2.20. M. Gryzinski and J.A. Kunc, *J. Phys. B: At. Mol. Phys.* **19** (1986) 2479.

2.21. Y.-K. Kim and M.E. Rudd, *Phys. Rev. A* **50** (1994) 3954.

2.22. S.M. Younger, *J. Quant. Spectr. Rad. Transfer* **26** (1981) 329; *Phys. Rev. A* **22** (1980) 111.

2.23. I. Bray and A.T. Stelbovics, *Phys. Rev. Lett.* **70** (1993) 746.

2.24. D. Kato and S. Watanabe, *Phys. Rev. Lett.* **74** (1995) 2443.

2.25. K. Bartschat and I. Bray, *J. Phys. B: At. Mol. Opt. Phys.* **29** (1996) L577.

2.26. M.S. Pindzola and F. Robicheaux, *Phys. Rev. A* **54** (1996) 2142.

2.27. H.R. Griem, Y. V. Ralchenko and I. Bray, *Phys. Rev. E* **56** (1997) 7186.

2.28. S. Glenzer and H.-J. Kunze, *Phys. Rev. A* **53** (1996) 2225.

2.29. S. Alexiou, S. Glenzer and R.W. Lee, *Phys. Rev. E* **60** (1999) 6238.

2.30. J.H. Macek, S. Y. Ovchinnikov and S.V. Pasovets, *Phys. Rev. Lett.* **74** (1995) 4631.

2.31. K.G. Kay, *Phys. Rev. Lett.* **83** (1999) 5190.

2.32. I. Pomeranchuk and J. Smorodinskii, *J. Phys. USSR* **9** (1945) 97.

2.33. F.G. Werner and J.A. Wheeler, *Phys. Rev.* **109** (1958) 126.

2.34. D. Rein, *Z. Phys.* **221** (1969) 423.

2.35. V.S. Popov, *Sov. J. Nucl. Phys.* **12** (1971) 235.

2.36. M.E. Rose, *Relativistic Electron Theory* (Wiley, New York) 1961, Secs. 28, 29 and 39.

2.37. W. Greiner and J. Reinhardt, *Quantum Electrodynamics* (Springer, New York) 1994, Sec. 7.1.

2.38. W. Greiner, *Relativistic Quantum Mechanics, Wave Equations* (Springer, New York) 1994, Sec. 9.

2.39. S. Flügge, *Practical Quantum Mechanics* (Springer, New York) 1974, Chapter II.D.

2.40. L.D. Landau and E.M. Lifshitz, *Quantum Mechanics* (Pergamon, Oxford) 1965.

2.41. I. Bialynicki-Birula, M. Cieplak and J. Kaminski, *Theory of Quanta* (Oxford University Press, New York) 1992.

2.42. E. Schrödinger, *Ann. Phys.* **79** (1926) 361.

2.43. D. Zwillinger, *Standard Mathematical Tables and Formulae* (CRC Press, New York) 1996, p. 54.

2.44. S. Flügge, *Practical Quantum Mechanics* (Springer, New York) 1974, Problem 202.

2.45. V.B. Berestetskii, E.M. Lifshitz and P. Pitaevskii, *Relativistic Quantum Theory* (Pergamon, Oxford) 1971, Sec. 36.

2.46. G.G. Simon, C. Schmitt, F. Borkowski and V.H. Walther, *Nucl. Phys.* **A333** (1980) 381.

2.47. D.H. Perkins, *Introduction to High Energy Physics* (Addison-Wesley, Menlo Park) 1987, Sec. 6.5.

2.48. R.P. Feynman, *Photon–Hadron Interactions* (Addison-Wesley, Redwood City, CA) 1989, Lecture 24.

## Chapter 3

3.1. F.B. Rosmej and V.S. Lisitsa, *Phys. Lett.A* **244** (1988) 401.

3.2. R.C. Isler and R.E. Olson, *Phys. Rev. A* **37** (1988) 3399.

3.3. S.S. Churilov, L.A. Dorokhin, Y. V. Sidelnikov, K.N. Koshelev, A. Schulz and Y. V. Ralchenko, *Contributions to Plasma Physics* **40** (2000) 167.

3.4. R.C. Elton, *X-Ray Lasers* (Academic Press, New York) 1990.

3.5. F.I. Bunkin, V.I. Derzhiev and S.I. Yakovlenko, *Sov. J. Quant. Electronics* **11** (1981) 981.

3.6. A.V. Vinogradov and I.I. Sobelman, *Sov. Phys. JETP* **36** (1973) 1115.

3.7. J. Von Neumann and E. Wigner, *Phys. Z.* **30** (1929) 467.

3.8. S.S. Gershtein and V.D. Krivchenkov, *Sov. Phys. JETP* **13** (1961) 1044.

3.9. L.I. Ponomarev and T.P. Puzynina, *Sov. Phys. JETP* **25** (1967) 846.

3.10. J.D. Power, *Phil. Trans. Roy. Soc. London* **A274** (1973) 663.

3.11. I.V. Komarov, L.I. Ponomarev and S. Yu. Slavyanov, *Spheroidal and Coulomb Spheroidal Functions* (Nauka, Moscow) 1976. [in Russian]

3.12. St. Böddeker, H.-J. Kunze, and E. Oks, *Phys. Rev. Lett.* **75** (1995) 4740.

3.13. E. Oks and E. Leboucher-Dalimier, *Phys. Rev. E, Rapid Commun.* **62** (2000) R3067.

3.14. E. Oks and E. Leboucher-Dalimier, *J. Phys. B* **33** (2000) 3795.

3.15. E. Leboucher-Dalimier, E. Oks, E. Dufour, P. Sauvan, P. Angelo, R. Schott and A. Poquerusse, *Phys. Rev. E, Rapid Commun.* **64** (2001) 065401.

3.16. E. Leboucher-Dalimier, E. Oks, E. Dufour, P. Angelo, P. Sauvan, R. Schott and A. Poquerusse, *Eur. Phys. J. D* **20** (2002) 269.

3.17. E. Dalimier, E. Oks, O. Renner and R. Schott, *J. Phys. B* **40** (2007) 909.

3.18. E. Oks, *Phys. Rev. Lett.* **85** (2000) 2084.

3.19. E. Oks, *J. Phys. B: At. Mol. Opt. Phys.* **33** (2000) 3319.

3.20. M.R. Flannery and E. Oks, *Phys. Rev. A* **73** (2006) 013405.

3.21. E. Oks, *Phys. Rev. E* **63** (2001) 057401.

3.22. N. Kryukov and E. Oks, *Int. Rev. At. Mol. Phys.* **2** (2011) 57.

3.23. A.P. Mishra, T. Nandi, and B.N. Jagatap, *J. Quant. Spectr. Rad. Transfer* **120** (2013) 114.

3.24. N. Kryukov and E. Oks, *Can. J. Phys.* **90** (2012) 647.

3.25. M.R. Flannery and E. Oks, *Eur. Phys. J. D* **47** (2008) 27.

3.26. G. Nogues, A. Lupascu, A. Emmert, M. Brune, J.-M. Raimond and S. Haroche, in *Atom Chips*, eds. J. Reichel and V. Vuletic (Wiley-VCH, Weinheim, Germany) 2011, Chapter 10, Sec. 10.3.3.

3.27. J.N. Tan, S.M. Brewer and N.D. Guise, *Phys. Scripta* **T144** (2011) 014009.

3.28. N. Kryukov and E. Oks, *Int. Rev. At. Mol. Phys.* **3** (2012) 17.

3.29. J.S. Dehesa, S. Lopez-Rosa, A. Martinez-Finkelshtein and R.J. Janez, *Int. J. Quantum Chemistry* **110** (2010) 1529.

3.30. T. Nandi, *J. Phys. B: At. Mol. Opt. Phys.* **42** (2009) 125201.

3.31. U.D. Jentschura, P.J. Mohr, J.N. Tan and B.J. Wundt, *Phys. Rev. Lett.* **100** (2008) 160404.

3.32. A.V. Shytov, M.I. Katsnelson and L.S. Levitov, *Phys. Rev. Lett.* **99** (2007) 246802.

3.33. M. Devoret, S. Girvin and R. Schoelkopf, *Ann. Phys.* **16** (2007) 767.

3.34. E. Oks, *Eur. Phys. J. D* **28** (2004) 171.

3.35. L. Holmlid, *J. Phys.: Condensed Matter* **14** (2002) 13469.

3.36. S.K. Dutta, D. Feldbaum, A. Walz-Flannigan, J.R. Guest and G. Raithel, *Phys. Rev. Lett.* **86** (2001) 3993.

3.37. H. Carlsen and O. Goscinski, *Phys. Rev. A* **59** (1999) 1063.

3.38. M. Day and M. Ebel, *Phys. Rev. B* **19** (1979) 3434.

3.39. D.J. Pegg, P.M. Griffin, B.M. Johnson, K.W. Jones, J.L. Cecchi and T.H. Kruse, *Phys. Rev. A* **16** (1977) 2008.

3.40. L.D. Landau and E.M. Lifshitz, *Mechanics* (Pergamon, Oxford) 1960.

3.41. H. Goldstein, *Classical Mechanics* (Addison-Wesley, Reading, MA) 1980.

3.42. L.D. Landau and E.M. Lifshitz, *Quantum Mechanics* (Pergamon, Oxford) 1965.

3.43. N. Kryukov and E. Oks, *Phys. Rev. A* **85** (2012) 054503.

3.44. M. Born and R. Oppenheimer, *Ann. Phys.* **84** (1927) 457.

3.45. W.R.S. Garton and F.S. Tomkins, *Astrophys. J.* **158** (1969) 839.

3.46. M. Brack and R.K. Bhaduri, *Semiclassical Physics* (Addison-Wesley, Reading, MA) 1997, Sec. 1.4.2.

3.47. W. Pauli, *Ann. Phys.* **68** (1922) 177.

3.48. D. Salzmann, *Atomic Physics in Hot Plasmas* (Oxford University Press, Oxford) 1998, Chapters 2 and 3.

3.49. M.S. Murillo and J.C. Weisheit, *Phys. Rep.* **302** (1998) 1.

3.50. H.R. Griem, *Principles of Plasma Spectroscopy* (Cambridge University Press, Cambridge) 1997, Secs. 5.5 and 7.3.

3.51. R.P. Drake, *High-Energy-Density-Physics: Fundamentals, Inertial Fusion, and Experimental Astrophysics* (Springer, Berlin) 2006, Sec. 3.2.2.

3.52. S. Atzeni and J. Meyer-ter-Vehn, *The Physics of Inertial Fusion: Beam Plasma Interaction, Hydrodynamics, Hot Dense Matter* (Oxford University Press, New York) 2004, Sec. 10.1.4.

3.53. J. Stein, I.B. Goldberg, D. Shalitin and D. Salzmann, *Phys. Rev. A* **39** (1989) 2078.

3.54. D. Salzmann, J. Stein, I.B. Goldberg and R.H. Pratt, *Phys. Rev. A* **44** (1991) 1270.

3.55. J. Stein and D. Salzmann, *Phys. Rev. A* **45** (1992) 3943.

3.56. P. Malnoult, B. D'Etat and H. Nguen, *Phys. Rev. A* **40** (1989) 1983.
3.57. Y. Furutani, K. Ohashi, M. Shimizu and A. Fukuyama, *J. Phys. Soc. Jpn.* **62** (1993) 3413.
3.58. P. Sauvan, E. Leboucher-Dalimier, P. Angelo, H. Derfoul, T. Ceccotti, A. Poquerusse, A. Calisti and B. Talin, *J. Quant. Spectrosc. Rad. Transfer* **65** (2000) 511.
3.59. H. Hora, *Plasmas at High Temperature and Density* (Springer, Berlin) 1991, Sec. 12.4.
3.60. Laser Interaction with Matter, *Proc. 23rd Eur. Conf.*, Oxford, 1994, ed. S.J. Rose, Inst. of Physics Conference Series No. 140 (IOP, Bristol and Philadelphia) 1995, pp. 173, 215, 221.
3.61. M.A. Liberman, J.S. De Groot, A. Toor and R.B. Spielman, *Physics of High-Density Z-Pinch Plasmas* (Springer, New York) 1999.

## Chapter 4

4.1. P.A.M. Dirac, *Can. J. Math.* **2** (1950) 129.
4.2. P.A.M. Dirac, *Proc. Roy. Soc. (London)* **A246** (1958) 326.
4.3. P.A.M. Dirac, *Lectures on Quantum Mechanics* (Academic Press, New York) 1964, Lecture 1.
4.4. P.A.M. Dirac, *Lectures on Quantum Field Theory* (Academic Press, New York) 1966, Lecture 14.
4.5. L.D. Landau and E.M. Lifshitz, *Mechanics* (Pergamon, Oxford) 1960.
4.6. L.L. Loeb, *Atomic Structure* (Wiley & Sons, London) 1938, Chapter VI.2.
4.7. L.D. Landau and E.M. Lifshitz, *Classical Theory of Fields* (Pergamon, Oxford) 1960.
4.8. H. Bacry, H. Ruegg and J.M. Souriau, *Commun. Math. Phys.* **3** (1966) 323.
4.9. D.M. Fradkin, *Progr. Theor. Phys.* **37** (1967) 798.
4.10. N. Mukunda, *Phys. Rev. (Second Series)* **155** (1967) 1383.
4.11. J.C. Tully and R.K. Preston, *J. Chem. Phys.* **55** (1971) 562; J.C. Tully, *J. Chem. Phys.* **93** (1990) 1061.

4.12. N.C. Blais and D.G. Truhlar, *J. Chem. Phys.* **79** (1983) 1334; N.C. Blais, D.G. Truhlar and C.A. Mead, *J. Chem. Phys.* **89** (1988) 6204.

4.13. M.F. Herman, *J. Chem. Phys.* **81** (1984) 754.

4.14. P.J. Kuntz, *J. Chem. Phys.* **95** (1991) 141; **95** (1991) 156.

4.15. E.J. Heller and R.C. Brown, *J. Chem. Phys.* **79** (1983) 3336.

4.16. P. Pechukas, *Phys. Rev.* **181** (1969) 166; **181** (1969) 174.

4.17. W.H. Miller and T.F. George, *J. Chem. Phys.* **56** (1972) 5637.

4.18. F.J. Webster, P.J. Rossky and R.A. Friesner, *Comput. Phys. Commun.* **63** (1991) 494.

4.19. D.A. Misha, *J. Chem. Phys.* **78** (1983) 7138.

4.20. D.J. Diestler, *J. Chem. Phys.* **78** (1983) 2240; L.L. Halcomb and D. J. Diestler, *J. Chem. Phys.* **84** (1986) 3130; M. Amarouche, F.X. Gadea and J. Durup, *Chem. Phys.* **130** (1989) 145.

4.21. H.-D. Meyer and W.H. Miller, *J. Chem. Phys.* **70** (1979) 3214.

4.22. H.-D. Meyer and W.H. Miller, *J. Chem. Phys.* **71** (1979) 2156; S.K. Gray and W.H. Miller, *Chem. Phys. Lett.* **93** (1982) 341.

4.23. G. Stock and W.H. Miller, *J. Chem. Phys.* **99** (1993) 1545.

4.24. J. Josephson, *Found. Phys.* **10** (1980) 243.

4.25. A.A. Sokolov, I.M. Ternov and V.C. Zhukovskii, *Quantum Mechanics* (Mir, Moscow) 1984.

4.26. L.I. Schiff, *Quantum Mechanics*, International Pure and Applied Physics Series (McGraw- Hill, New York) 1968.

4.27. A. Capri, *Relativistic Quantum Mechanics and Introduction to Quantum Field Theory* (World Scientific, Singapore) 2002.

4.28. A. Einstein, *Ann. Phys.* **49** (1916) 769.

4.29. J.D. Walecka, *Introduction to General Relativity* (World Scientific, Singapore) 2007.

4.30. S. Schutz, *A First Course in General Relativity* (Cambridge University Press, New York) 1985.

4.31. B. M. Karnakov, Ph. A. Korneev and S. V. Popruzhenko, *J. Exp. Theor. Phys.* **106** (2008) 650.

4.32. A. Holas and N.H. March, *J. Phys. A: Math. Gen.* **23** (1990) 735.

4.33. J.A. Camarena and E. Oks, *Int. Rev. At. Mol. Phys.* **1** (2010) 143.

## Chapter 5

5.1. H. Brauning, H. Helm, J.S. Briggs and E. Salzborn, *Phys. Rev. Lett.* **99** (2007) 173202.

5.2. G.V. Sholin, A.E. Trenin, V.A. Belyaev, M.M. Dubrovin and A.A. Terent'ev, *Zh. Eksp. Teor. Fiz.* **131** (2007) 228 [*JETP* **104** (2007) 201].

5.3. F. Brouillard, W. Claeys, G. Poulaert, G. Rahmat and C. Van Wassenhove, *J. Phys. B: At. Mol. Phys.* **12** (1979) 1253.

5.4. V.A. Belyaev, M.M. Dubrovin, A.A. Teren'tev, A.E. Trenin and G.V. Sholin, *Fiz. Plazmy* **27** (2001) 1093 [*Plasma Phys. Rep.* **27** (2001) 1032].

5.5. V.A. Belyaev, M.M. Dubrovin, A.A. Teren'tev, A.E. Trenin and G.V. Sholin, *Fiz. Plazmy* **27** (2001) 955 [*Plasma Phys. Rep.* **27** (2001) 901].

5.6. H.A. Bethe, in *Handbuch der Physik*, ed. S. Flügge, **v. 24-1** (Springer, Berlin) 1933; (ONTI, Leningrad) 1935. [in German and in Russian]

5.7. H.A. Bethe and E.E. Salpeter, *Quantum Mechanics of One- and Two-Electron Atoms* (Springer, New York) 1957; (Fizmatgiz, Moscow) 1960.

5.8. A.M. El'yashevich, *Atomic and Molecular Spectroscopy* (Fizmatgiz, Moscow) 1962. [in Russian]

5.9. L. Gaunt, *Proc. R. Soc. London, Ser. A* **122** (1929) 153.

5.10. W. Heisenberg, *Z. Phys.* **39** (1927) 499.

5.11. V.S. Senashenko and U.I. Safronova, *The Theory of Spectra of Multicharged Ions* (Energoatomizdat, Moscow) 1984. [in Russian]

5.12. L.D. Landau and E.M. Lifshitz, *Quantum Mechanics* (Nauka, Moscow) 1989; (Butterworth–Heinemann, Oxford) 1991.

5.13. S.V. Vonsovskii, *Magnetism* (Nauka, Moscow) 1971; (Wiley, New York) 1974.

5.14. V.M. Galitskii, B.M. Karnakov and V.I. Kogan, *Problems in Quantum Mechanics* (Nauka, Moscow) 1981. [in Russian]

5.15. S.E. Moore, *Ionization Potentials and Ionization Limits Derived from the Analyses of Optical Spectra* (National Standard Reference Data Series — National Bureau of Standards (NSRDS–NBS), Washington) **v. 34**, 1970.

## Chapter 6

6.1. D.R. Inglis and E. Teller, *Astrophys. J.* **90** (1939) 439.

6.2. J. Holtsmark, *Ann. Phys.* **58** (1919) 577.

6.3. J. Hey, *J. Phys. B: At. Mol. Opt. Phys.* **46** (2013) 175702.

6.4. H.R. Griem, *Plasma* Spectroscopy (McGraw-Hill, New York) 1964.

6.5. B.L. Welch, H.R. Grim, J. Terry, C. Kurz, B. LaBombard, B. Lipschultz, E. Maramar and G. McCracken, *Phys. Plasmas* **2** (1995) 4246.

6.6. F.D. Rosenberg, U. Feldman and G.A. Doschek, *Astrophys. J.* **212** (1977) 905.

6.7. U. Feldman and G.A. Doschek, *Astrophys. J.* **212** (1977) 913.

6.8. E. Oks, *Int. Rev. At. Mol. Phys.* **4**, (2013) 105; *J. Quant. Spectr. Rad. Transfer* **156** (2015) 24.

6.9. M. Born, *The Mechanics of the Atom* (Bell and Sons, London) 1927.

6.10. Y.N. Demkov, B.S. Monozon and V.N. Ostrovsii, *Sov. Phys. JETP* **30** (1970) 775.

## Chapter 7

7.1. K.M. Sando, R.O. Doyle and A. Dalgarno, *Astrophys. J.* **157** (1969) L143.

7.2. J.C. Stewart, J.M. Peek and J. Cooper, *Astrophys. J.* **179** (1973) 983.

7.3. J.L. Queffelec and M. Girault, C.R. *Acad. Sci.* Paris **B279** (1974) 649.

7.4. Le Quang Rang and D. Voslamber, *J. Phys. B* **8** (1975) 331.

7.5. R.C. Preston, *J. Phys. B.* **10** (1977) 523.

7.6. N. Allard and J. Kielkopf, *Rev. Mod. Phys.* **54** (1982) 1103.

7.7. E. Nelan and G. Wegner, *Astrophys. J.* **289** (1985) L31.

7.8. Ph. Malnoult, B. d'Etat and H. Nguyen, *Phys. Rev. A* **40** (1989) 1983.

7.9. E. Leboucher-Dalimier, A. Poquerusse, P. Angelo, I. Gharbi and H. Derfoul, *J. Quant. Spectr. Rad. Transfer.* **51** (1994) 187.

7.10. N.F. Allard, J. Kielkopd and N. Feautrier, Astron. *Astrophys.* **330** (1998) 782.

7.11. N.F. Allard, I. Drira, M. Gerbaldi, J. Kielkopf and A. Spielfiedel, Astron. *Astrophys.* **335** (1998) 1124.

7.12. E. Oks and E. Leboucher-Dalimier, *Phys. Rev. E. Rapid Communications* **62** (2000) R3067.

7.13. E. Oks, *Plasma Spectroscopy: The Influence of Microwave and Laser Fields, Springer Series on Atoms and Plasmas*, **v. 9** (Springer, New York) 1995.

7.14. J. von Neumann and E. Wigner, *Phys. Z* **30** (1929) 467.

7.15. S.S. Gershtein and V.D. Krivchenkov, *Sov. Phys. JETP.* **13** (1961) 1044.

7.16. N. Kryukov and E. Oks, *Phys. Rev. A* **85** (2012) 054503.

7.17. J.D. Power, *Phil. Trans. Roy. Soc. London.* **A274** (1973) 663.

7.18. I.V. Komarov, L.I. Ponomarev and S.Yu. Slavyanov, *Spheroidal and Coulomb Spheroidal Functions* (Nauka, Moscow) 1976 [in Russian].

7.19. A.B. Underhill and J.H. Waddell, *Nat. Bur. Stand.* (USA) *Circ.* **603** (1959).

7.20. L.D. Landau and E.M. Lifshitz, *Classical Theory of Fields* (Pergamon, Oxford) 1962.

7.21. I.I. Sobel'man, *Introduction to the Theory of Atomic Spectra* (Pergamon, Oxford) 1972, Sec. 36.3.

7.22. E. Oks, St. Böddeker and H.-J. Kunze, *Phys. Rev. A* **44** (1991) 8338.

7.23. E. Oks, *Stark Broadening of Hydrogen and Hydrogenlike Spectral Lines in Plasmas: The Physical Insight* (Alpha Science International: Oxford, United Kingdom) 2006.

7.24. E. Leboucher-Dalimier, E. Oks, E. Dufour, P. Sauvan, P. Angelo, R. Schott and A. Poquerusse, *Phys. Rev. E*, Rapid Communications **64** (2001) 065401.

7.25. E. Dalimier, E. Oks, O. Renner and R. Schott, *J. Phys. B: Atom. Mol. Opt. Phys.* **40** (2007) 909.

7.26. O. Renner, E. Dalimier, R. Liska, E. Oks and M. Šmíd, *J. Phys. Conf. Ser.* **397** (2012) 012017.

7.27. E. Oks, E. Dalimier, A. Ya. Faenov, T. Pikuz, Y. Fukuda, S. Jinno, H. Sakaki, H. Kotaki, A. Pirozhkov, Y. Hayashi, I. Skobelev, T. Kawachi, M. Kando and K. Kondo, *J. Phys. B: At. Mol. Opt. Phys.* Fast Track Communications, 2014.

7.28. E. Dalimier and E. Oks, *J. Phys. B: Atom. Mol. Opt. Phys.* **47** (2014) 105001.

7.29. Ju. N. Demkov, Trudy Gosudarstvennogo Opticheskogo Instituta (Proceedings of the State Optical Institute), 1978, 43, 71 (in Russian).

7.30. S.I. Nikitin and V.N. Ostrovsky, *J. Phys. B: Atom. Mol. Opt. Phys.* **9** (1976) 3141.

## Chapter 8

8.1. P.M. Koch and K.A.H. van Leeuwen, *Phys. Rep.* **255** (1995) 289.

8.2. B.I. Meerson, E. Oks and P. Sasorov, *JETP Lett.* **29** (1979) 72.

8.3. G. Casati, B.V. Chirikov, D.M. Shepelyansky and I. Guarneri, *Phys. Rep.* **154** (1987) 77.

8.4. H. Hasegawa, M. Robnik and G. Wunner, *Prog. Theor. Phys. Suppl.* **98** (1989) 198.

8.5. T.F. Gallagher, *Rydberg Atoms* (Cambridge University Press, Cambridge) 1994.

8.6. J.-P. Connerade, *Highly Excited Atoms* (Cambridge University Press, Cambridge) 1998.

8.6. E. Oks and T. Uzer, *J. Phys. B: At. Mol. Opt. Phys.* **32** (1999) 3601; **33** (2000) 1985; **33** (2000) 2207; **33** (2000) L253; **33** (2000) 5357.

8.7. B.B. Nadezhdin and E. Oks, *Sov. Phys. Tech. Phys. Lett.* **12** (1986) 512.

8.8. E. Oks, J.E. Davis and T. Uzer, *J. Phys. B: At. Mol. Opt. Phys.* **33** (2000) 207.

8.9. J.T. Wheeler, *Can. J. Phys.* **83** (2005) 91.

8.10. S. Boeddeker, H.-J. Kunze and E. Oks, *Phys. Rev. Lett.* **75** (1995) 4740.

8.11. E. Oks, *J. Phys. B: At. Mol. Opt. Phys.* **33** (2000) 3319.

8.12. E. Oks and E. Leboucher-Dalimier, *J. Phys. B: At. Mol. Opt. Phys.* **33** (2000) 3795.

8.13. E. Dalimier, E. Oks, O. Renner and R. Schott, *J. Phys. B: At. Mol. Opt. Phys.* **40** (2007) 909.

8.14. E. Leboucher-Dalimier, E. Oks, E. Dufour, P. Sauvan, P. Angelo, R. Schott and A. Poquerusse, *Phys. Rev. E, Rapid Commun.* **64** (2001) 065401.

8.15. O. Renner, E. Dalimier, R. Liska, E. Oks and M. Šmíd, *J. Phys. Conf. Ser.* **397** (2012) 012017.

8.16. E. Dalimier and E. Oks, *Int. Rev. At. Molec. Phys.* **3** (2012) 85.

8.17. E. Dalimier and E. Oks, *J. Phys. B: At. Mol. Opt. Phys.* **47** (2014) 105001.

8.18. E. Dalimier, E. Oks and O. Renner, *Atoms* **2** (2014) 178.

## Appendix A

A.1. L.I. Ponomarev, *Contemp. Phys.* **31** (1990) 219.

A.2. K. Nagamine, *Hyperfine Interactions* **138** (2001) 5.

A.3. K. Nagamine and L.I. Ponomarev, *Nucl. Phys. A* **721** (2003) C863.

A.4. C. Chelkowsky, A.D. Bandrauk and P.B. Corkum, *Laser Phys.* **14** (2004) 473.

A.5. J. Guffin, G. Nixon, D. Javorsek II, S. Colafrancesco and E. Fischbach, *Phys. Rev. D* **66** (2002) 123508.

A.6. A.R.P. Rau, *J. Astrophys. Astr.* **17** (1996) 113.

A.7. P. Balling, H.H. Andersen, C.A. Brodie, U.V. Pedersen, V.V. Petrunin, M.K. Raarup, P. Steiner and T. Andersen, *Phys. Rev. A* **61** (2000) 022702.

A.8. K. Sakimoto, *Phys. Rev. A* **78** (2008) 042509.

A.9. D.F. Measday, *Phys. Reports* **354** (2001) 243.

A.10. K. Sakimoto, *Phys. Rev. A* **81** (2010) 012511.

A.11. J.D. Garcia, N.H. Kwong and J.S. Cohen, *Phys. Rev. A* **35** (1987) 4068.

A.12. N. Kryukov and E. Oks, *Can. J. Phys.* (2014) 10.1139/cjp-2013-0705.

## Appendix B

B.1. G.R. Fowles and G.L. Cassiday, *Analytical Mechanics* (Thomson Brooks/Cole, Belmont, CA) 2005, Sec. 3.6.

B.2. S.T. Thornton and J.B. Marion, *Classical Dynamics of Particles and Systems* (Thomson Brooks/Cole, Belmont, CA) 2004, Sec. 3.6.

B.3. D.I. Blochinzew, *Phys. Z. Sow. Union* **4** (1933) 501.

B.4. E. Oks, *Plasma Spectroscopy. The Influence of Microwave and Laser Fields* (Springer, New York) 1995, Sec. 3.1.

Printed in the United States
By Bookmasters